14. Band, 3. Heft

Fortschritte der chemischen Forschung
Topics in Current Chemistry

Herausgeber:

Prof. Dr. *A. Davison*   Department of Chemistry, Massachusetts Institute
of Technology, Cambridge, MA 02139, USA

Prof. Dr. *M. J. S. Dewar*   Department of Chemistry, The University of Texas
Austin, TX 78712, USA

Prof. Dr. *K. Hafner*   Institut für Organische Chemie der TH
6100 Darmstadt, Schloßgartenstraße 2

Prof. Dr. *E. Heilbronner*   Physikalisch-Chemisches Institut der Universität
CH-4000 Basel, Klingelbergstraße 80

Prof. Dr. *U. Hofmann*   Institut für Anorganische Chemie der Universität
6900 Heidelberg 1, Tiergartenstraße

Prof. Dr. *K. Niedenzu*   University of Kentucky, College of Arts and Sciences
Department of Chemistry, Lexington, KY 40506, USA

Prof. Dr. *Kl. Schäfer*   Institut für Physikalische Chemie der Universität
6900 Heidelberg 1, Tiergartenstraße

Prof. Dr. *G. Wittig*   Institut für Organische Chemie der Universität
6900 Heidelberg 1, Tiergartenstraße

Schriftleitung:

Dipl.-Chem. *F. Boschke*   Springer-Verlag, 6900 Heidelberg 1, Postfach 1780

Springer-Verlag   6900 Heidelberg 1 · Postfach 1780
Telefon (06221) 49101 · Telex 04-61723
1000 Berlin 33 · Heidelberger Platz 3
Telefon (0311) 822001 · Telex 01-83319

Springer-Verlag   New York, NY 10010 · 175, Fifth Avenue
New York Inc.   Telefon 673-2660

ISBN 978-3-540-04818-3   ISBN 978-3-540-36197-8 (eBook)
DOI 10.1007/978-3-540-36197-8

Titel-Nr. 7720

Acylierung von Enaminen

Prof. Dr. S. Hünig und Dr. H. Hoch

Institut für Organische Chemie der Universität Würzburg

Inhalt

S. Hünig und H. Hoch

1. Einführung

1.1 Abgrenzung des Themas

Als Enamine sind alle Verbindungen zu bezeichnen, in denen eine Aminogruppe unmittelbar mit einer Doppelbindung verknüpft ist. Hier soll
nur von Enaminen die Rede sein, deren N-Alkyl- oder Alkylaryl-Gruppen
eine Tautomerie ausschließen und deren Doppelbindung keine Substituenten mit starker Rückwirkung aufweisen. Typische Vertreter sind
1, 2 und *3*. Auch *4*, die sogenannte Fischerbase[1], ist als Enamin zu be-
trachten. Ja, man muß sagen, daß alle wichtigen Alkylierungs- und
Acylierungsreaktionen der Enamine *1—3*, soweit sie zu offenkettigen
Produkten führen, schon von der Fischerbase *4* bekannt waren[1]. Ihre
Übertragung auf Enamine vom Typ *2*, über den zuerst G. *Stork* u. Mit-
arb.[2,3] berichteten, führte, nach Ausdehnung auf *1*, zu einer Fülle von
präparativ wichtigen und theoretisch interessanten Reaktionen, über
die inzwischen mehrere Zusammenfassungen erschienen sind [4a—e].

Der vorliegende Überblick beschränkt sich auf die Acylierung der
Enamine durch reaktive Derivate der Carbonsäuren und der Kohlen-
säure, auch wenn Cycloaddukte als Zwischen- oder Endprodukte auftreten. Die Reaktion mit Sulfonsäure-Derivaten bleibt unberücksich-
tigt[5]. Auf das ähnliche Verhalten der Ketenaminale *5a* Keten-O,N-acetale
5b und Ketenacetale *5c* bei der Acylierung sei lediglich hingewiesen[6]. Vor-
angeschickt werden die wichtigsten Syntheseprinzipien für Enamine.
Außerdem ist es zweckmäßig, vorab das Verhalten der Enamine gegenüber Protonen zu besprechen, die bei Acylierungsreaktionen kaum auszuschließen sind. Die ebenfalls wichtige hydrolytische Spaltung der

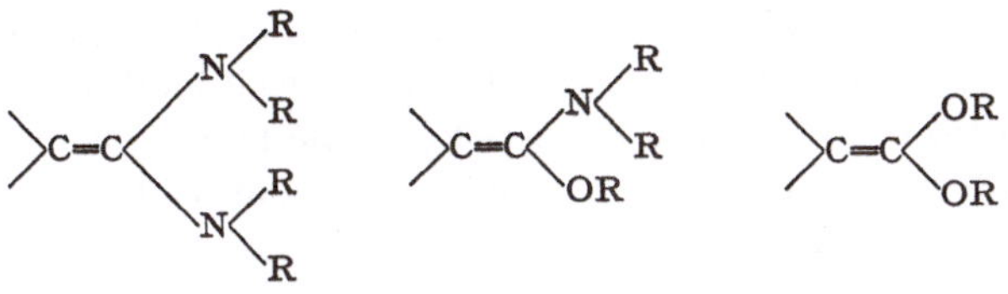

$$5a \qquad\qquad 5b \qquad\qquad 5c$$

Enamine und ihrer Acylierungsprodukte erscheint bei den einzelnen Beispielen.

1.2. Die wichtigsten Syntheseprinzipien für Enamine

Erst die glatte Synthese von Enaminen *aus Aldehyden* nach *Mannich* und *Davidsen*[7] (Gl. 1) sowie *aus Ketonen* nach *Herr* und *Heyl*[8] (Gl. 2) mit sec. Aminen wie Dimethyl- und Diäthylamin und vor allem Pyrrolidin, Piperidin und Morpholin schaffte die Voraussetzung für die allgemeine Anwendbarkeit der Enamine.

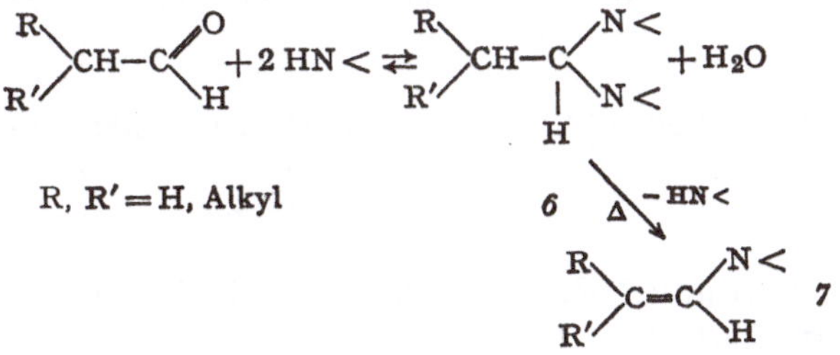

R, R′ = H, Alkyl

(1)

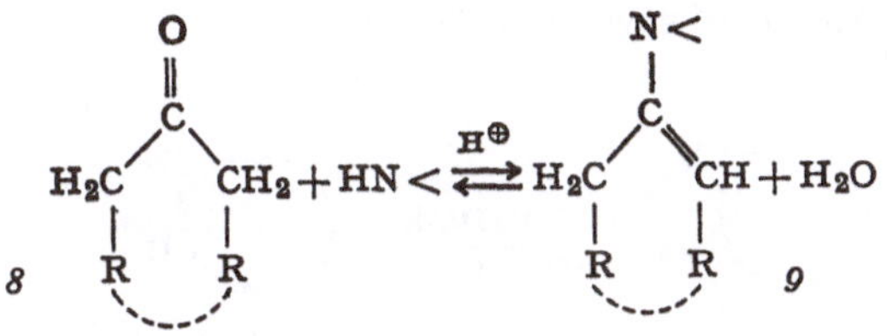

(2)

Der Wasserentzug geschieht im ersten Falle mit Kaliumcarbonat[7], welches zugleich Säurespuren entfernt, die sonst leicht zur Selbstkondensation des Aldehyds führen könnten (s. u.). Das zunächst entstehende Aminal *6* liefert bei der destillativen Spaltung das Enamin *7*[7]. Während dieses Verfahren mit Ketonen *8* nur unbefriedigend verläuft[7], gelingt die Synthese der Enamine *9* aus *8* und sec. Amin glatt, wenn unter

Säurekatalyse das Wasser azeotrop mit Benzol oder Toluol entfernt wird[8]. Insbesondere cyclische Ketone eignen sich hierfür.

Der kritische Schritt der Wasserabspaltung kann umgangen werden. So läßt sich das sonst schwer zugängliche Enamin *11* aus dem Ketal *10* und N-Methylanilin gewinnen[9].

$$CH_3-C{\overset{\displaystyle OC_2H_5}{\underset{\displaystyle C_6H_5}{\big|}}}OC_2H_5 \quad \xrightarrow{\ C_6H_5NCH_3\ \overset{H}{|}\ } \quad CH_2=C{\underset{\displaystyle C_6H_5}{\big|}}-N{\overset{\displaystyle CH_3}{\underset{\displaystyle C_6H_5}{\big\langle}}}$$

10   *11*

Die *Tris-dialkylaminoborane*[10] und *-arsine*[11], *12* bzw. *13*, verwandeln

$$(R_2N)_3B \qquad\qquad (R_2N)_3As$$

12   *13*

Aldehyde und Ketone glatt in Enamine. Ähnlich wirkt Titantetrachlorid bei großem Überschuß an sec. Amin, wie die folgende Gleichung (Gl. 3) zeigt[12a].

$$2\,RCH_2-COR^1 + 6\,NHR_2^2 + TiCl_4 \rightarrow 2\,RCH=C{\underset{\displaystyle NR_2^2}{\big|}}-R^1 + 4\,R_2^2NH_2Cl + TiO_2 \qquad (3)$$

Aldehyde wurden mit Tris-[dimethylamino]-methan in trans-Enamine übergeführt[12b].

Heterocyclische Enamine *15* werden gewöhnlich durch Dehydrierung der gesättigten Amine *14* gewonnen[13] (Gl. 4).

$$\left[\begin{array}{c} -CH_2 \\ | \\ CH_2 \\ N \\ | \\ R \end{array}\right] \quad \xrightarrow[\text{2) Base}]{\text{1) Hg(OAc)}_2} \quad \left[\begin{array}{c} -CH \\ \| \\ CH \\ N \\ | \\ R \end{array}\right] \qquad (4)$$

14 R   *15* R

1.3. Protonierung der Enamine und Folgereaktionen

Die Delokalisierung des unverbundenen Elektronenpaares am N-Atom, die sich im UV-spektroskopischen Verhalten der Enamine ausprägt, läßt sich durch die Grenzstrukturen *16a* und *16b* beschreiben.

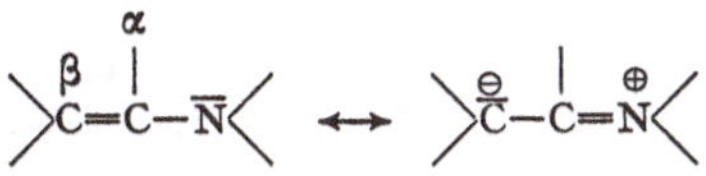

$$\overset{\alpha}{\underset{16a}{\underset{\beta}{\diagup}}}\text{C}=\overset{|}{\text{C}}-\bar{\text{N}}\diagdown \quad\longleftrightarrow\quad \underset{16b}{\diagup}\overset{\ominus}{\text{C}}-\overset{|}{\text{C}}=\overset{\oplus}{\text{N}}\diagdown$$

Infolge der damit verbundenen Nucleophilie des β-C-Atoms greifen Elektrophile zumeist in β-Stellung an — darauf beruht der präparative Wert der Enaminreaktionen — der Angriff am N-Atom wird ebenfalls beobachtet [3,14,15,18]. Die Natur des Aminrestes wirkt sich stark auf die Elektrophilie des β-C-Atoms aus. Für die häufig verwendeten Aminreste des Pyrrolidins, Piperidins und Morpholins liegen Vergleichswerte vor.

Rel. Geschw. der Reaktion von	$(CH_3)_2C=CHNR_2$		
$R_2N =$	Morpholino	Piperidino	Pyrrolidino
mit Dimethylketen [16]	1	7	20
Diphenylketen [17]	1	—	1400
$H_3O^{\oplus}$ [18]	1	330	26000

Einen entsprechenden Gang zeigen die Vinyl-H[1]-NMR-Signale der Enamine vom Typ 17 (vgl. [20]). Auch hier kommt die gegensätzliche Position von Morpholin und Pyrrolidin, d.h. die besonders starke Donatorwirkung des letzteren zum Ausdruck.

Abhängigkeit der H_β-NMR-Signale vom Aminrest (δ in ppm in CCl_4) [19]

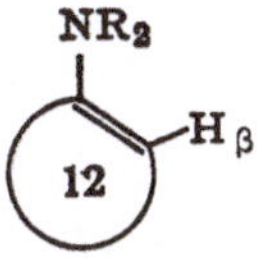

R_2N: Morpholino	Piperidino	Pyrrolidino
4.30	4.25	3.94

17

Bei der Protonierung hat sich 18 als kurzlebiges Produkt nachweisen lassen[18,21], doch kommt den stabilen Salzen zweifellos die Immonium-Struktur 19 zu, da das IR-Spektrum frei von N-H-Banden ist [22].

$$
16a \quad \text{>}C\text{=}C\text{-}\overset{|}{N}\text{<} \quad \underset{H^{\oplus}}{\overset{H^{\oplus}}{\rightleftharpoons}} \quad \text{>}C\text{=}C\text{-}\overset{H}{\underset{\oplus}{N}}\text{<} \quad 18
$$

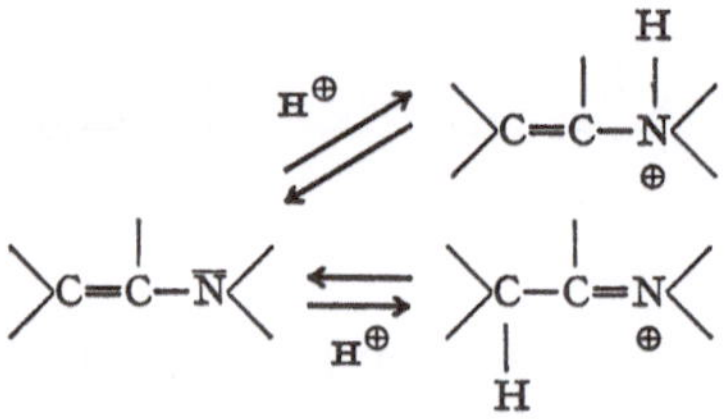

$$
\text{>}C\text{-}\overset{|}{\underset{H}{C}}\text{=}\overset{\oplus}{N}\text{<} \quad 19
$$

* IR: 1625—1690 cm^{-1} 1649—1705 cm^{-1}
  UV: 221— 235 nm (Hexan)   222— 233 nm (Acetonitril)
        $\varepsilon = 3500—9960$ $\varepsilon = 4100—19800$

* in Abhängigkeit vom Substituenten

Diese Zuordnung stützt der Vergleich mit Vinyl-ammonium-Salzen wie z.B. *20* und *21*, die zwischen 1600—1700 cm^{-1} keine Bande aufweisen, und deren UV-Absorption wesentlich kürzerwellig, im typischen Olefingebiet liegt [23,13].

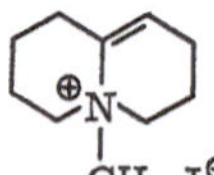

20

$$
CH_2\text{=}CH\text{-}\overset{CH_3}{\underset{CH_3}{\overset{\oplus}{N}}}\text{-}CH_3 \quad *21*
$$

$Br^{\ominus}$ [23)]

Es sei darauf hingewiesen, daß sich die optischen Daten von *16a* und *19* nicht signifikant unterscheiden. Erst bei sterischer Mesomeriebehinderung tritt N-Protonierung des Enamins ein, wie beim Neostrychnin (*22*) [24)] und ähnlichen Alkaloiden [25)] oder beim Δ²-Dehydro-chinuclidin (*23*) [26)]. Bereits die Basen weisen lediglich eine UV-Bande bei 213—215 nm auf, wie sie bei tert. aliphatischen Aminen auftritt.

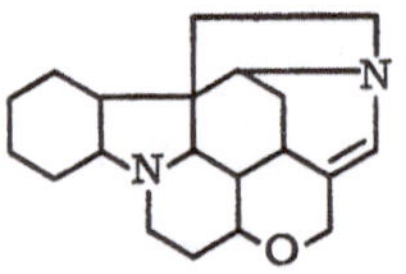

22.

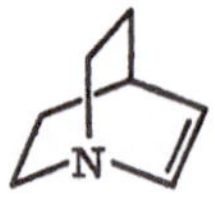

23

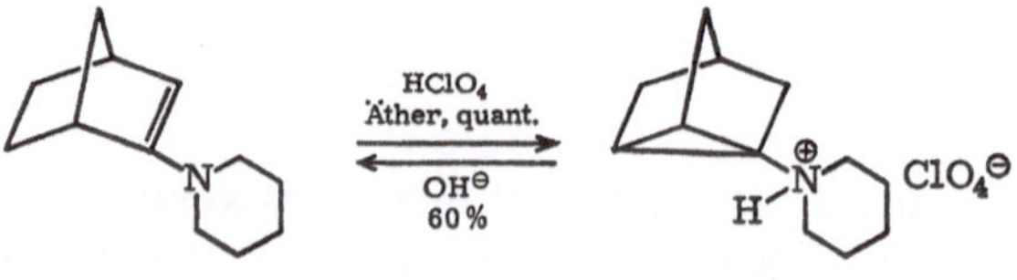

24   HClO$_4$ Äther, quant.   OH$^{\ominus}$ 60%   ClO$_4^{\ominus}$ *25*

Das gespannte 2-Piperidino-norbornen-(2) *24* verhält sich ebenfalls anormal; es wird zum Tricyclenderivat *25* protoniert [27]. Immonium-Salze vom Typ *19* sind außerordentlich hydrolyse-empfindlich [22]. Sie stellen zugleich die Zwischenstufe bei der besprochenen Enamin-Synthese dar, der protonenkatalysierten Kondensation von sec. Amin und Carbonylverbindung unter azeotropem Wasserentzug [8].

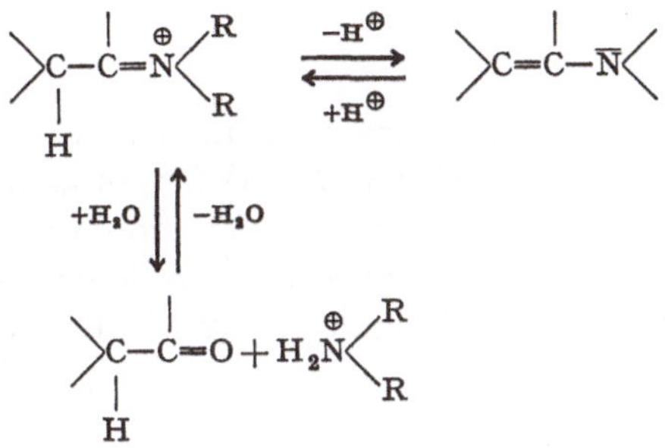

$$
\begin{array}{c}
\underset{H}{\overset{|}{>}}\!C\!-\!C\!=\!\overset{\oplus}{N}\!\!\diagdown^{R}_{R} \quad \overset{-H^{\oplus}}{\underset{+H^{\oplus}}{\rightleftharpoons}} \quad >\!C\!=\!C\!-\!\bar{N}\!< \\[3mm]
+H_2O \;\Big\updownarrow\; -H_2O \\[3mm]
\underset{H}{\overset{|}{>}}\!C\!-\!\overset{|}{C}\!=\!O + H_2\overset{\oplus}{N}\!\!\diagdown^{R}_{R}
\end{array}
\qquad (5)
$$

Nur Enamine mit heterocyclischem 5- oder 6-Ring, wie z.B. *26*, sind im wasserhaltigen Solvens reversibel protonierbar (z.B. zu *27* [28]), während höhergliedrige Ring-enamine (*28*) bereits hydrolysieren (→ *29*) [29].

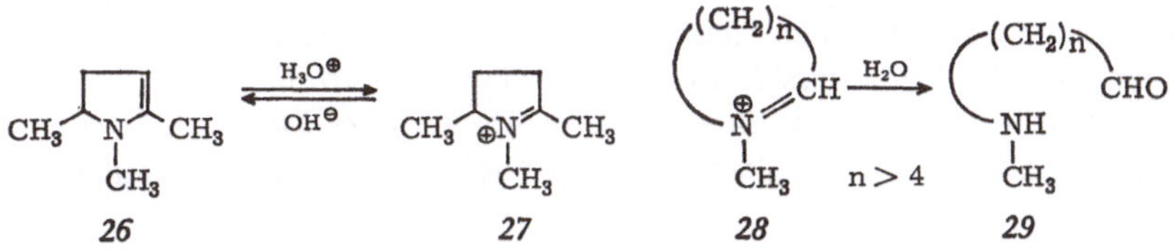

Enamine, bei denen sich nach der Hydrolyse Amin- und Carbonyl-funktion in getrennten Molekülen befinden, hydrolysieren schon durch Wasser allein [22,30]. Da die Reaktivität der Enamine mit der zugrunde-liegenden Carbonylverbindung zusammenhängt, wird im folgenden zwischen „*Aldehydenaminen*" und „*Ketonenaminen*" unterschieden.

Enamine, die zu glatt oximierbaren Carbonylverbindungen hydroly-sieren, sind leicht auf ihre Reinheit zu prüfen: Der Verbrauch an Säure durch freigesetztes Amin muß nach Zusatz von Hydroxylammonium-chlorid [31] der bei der Oximierung freigesetzten Säuremenge äquivalent sein [30].

Der glatten Protonierung der Enamine ist es zuzuschreiben, daß cis-Enamine vom Typ *30a* — durch Isomerisierung von N,N-Dialkylallyl-aminen mit Kalium-tert.-butylat in Dimethylsulfoxid gewonnen [32] —

nur bei rigorosem Ausschluß von Säurespuren zu fassen sind, da sie sonst augenblicklich in die stabilere trans-Form *30b* übergehen [32].

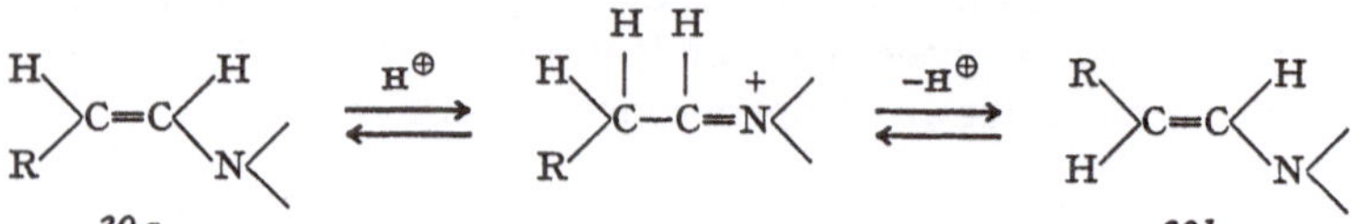

Wenn die Deprotonierung des Immoniumsalzes (z.B. *32*) zu verschiedenen Enaminen (z.B. *31* und *33* [33]) führen kann, werden grundsätzlich *beide Isomeren* beobachtet. Bei der protonenkatalysierten Synthese von Enaminen sowie bei Acylierungsreaktionen ist daher mit einer raschen Einstellung des Tautomeriegleichgewichtes zu rechnen [34,35].

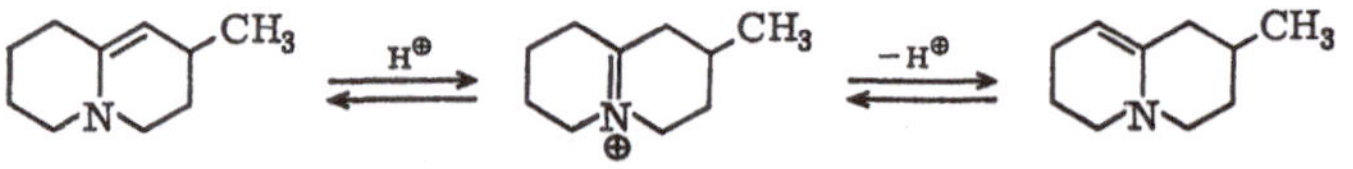

Die Lage des Gleichgewichtes hängt sowohl von elektronischen wie sterischen Faktoren ab [20]. Die Konstitution eines isolierten, tautomeriefähigen Enamins sagt daher nichts über die Verhältnisse im Reaktionsmedium aus [36]. Besonders Enamine vom Typ *34* unterliegen leicht Kondensationsreaktionen, die bereits beim Stehen oder Destillieren eintreten [37], und z.B. der Alkylierung mit Methyljodid den Rang ablaufen [38].

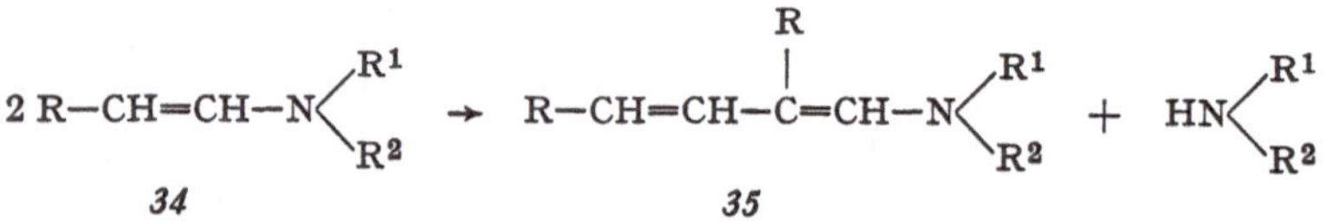

Selbst wenn man die Ketalmethode zur Enamin-Darstellung auf *36* überträgt, resultiert das Dienamin *37* als Folge einer Mannich-Kniss-

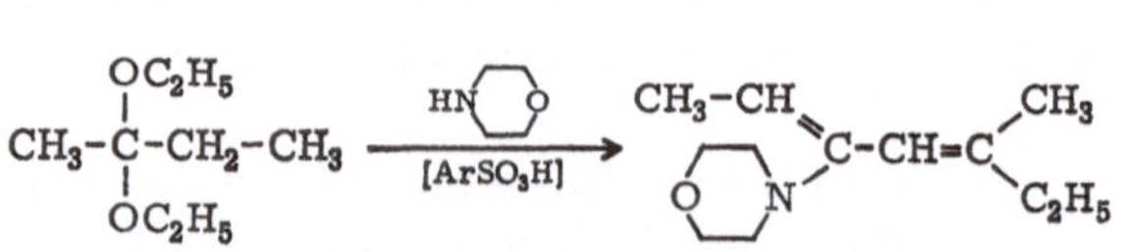

Kondensation [39]. Die Reaktion ist höchstwahrscheinlich säurekataly-siert, da sie bei Gegenwart der zugehörigen Immoniumionen sehr rasch eintritt [40] und ihr Verlauf sich an heterocyclischen Enaminen, bei denen die Abspaltung des sec. Amines unterbleibt, ablesen läßt. So ist das Enamin *42* bzw. sein Immoniumsalz *43* weder bei der Dehydrierung des N-Methyl-piperidins *38* [41] noch bei der Hydrierung von *39* [42] oder *40* [42] zu fassen. Es entsteht vielmehr stets in guter Ausbeute das Dimere *41*, das über *44* im Sinne einer reversiblen [43] Aldolkondensation von *42* und *43* entstanden sein muß.

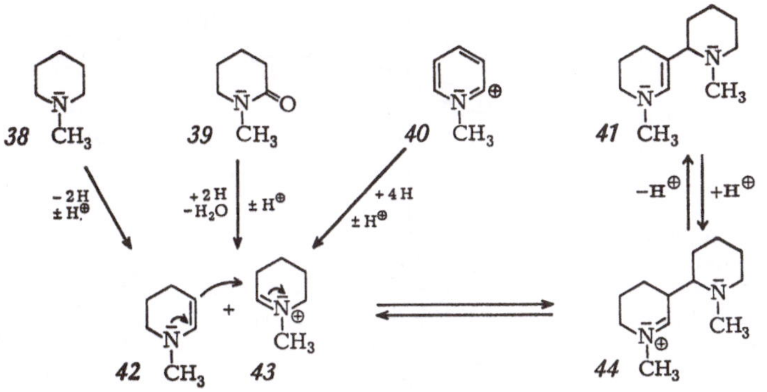

Diesem Reaktionstyp ist bei allen unter Protonenbeteiligung verlaufen-den Enaminreaktionen Beachtung zu schenken. Er macht darüberhinaus die altbekannte Dimerisierung von 2-Methylbenzthiazolium-Salzen beim Alkalisieren verständlich [43a]; auch die Bildung von Xantho- und Erythroapocyaninen aus Alkylchinolinium-Salzen und Alkalilauge [44,45] ist durch diesen Kondensationstyp zu deuten.

2. Acylierung mit Carbonsäurederivaten

2.1. Acylierung mit Ketenen

Dieser Reaktionstyp kann zu vielfältigen Produkten führen, die auch dann entstehen, wenn mit Carbonsäurechloriden acyliert wird, die unter den Reaktionsbedingungen in Ketene übergehen können.

Es werden hier daher alle Reaktionen mit Enaminen behandelt, bei denen in Substanz eingesetzte oder in situ erzeugte Ketene das Reak-tionsbild bestimmen [45a].

2.1.1. Acylierung von „Aldehydenaminen"

Dimethylketen (45) und das Enamin 46 a bzw. 46 b reagieren exotherm unter Bildung

des stabilen 1:1-Cycloadduktes 49 [46],
des 2:1-Adduktes 48 [47,48]
und einer geringen Menge eines bisher unbekannten 3:1-Adduktes [46].

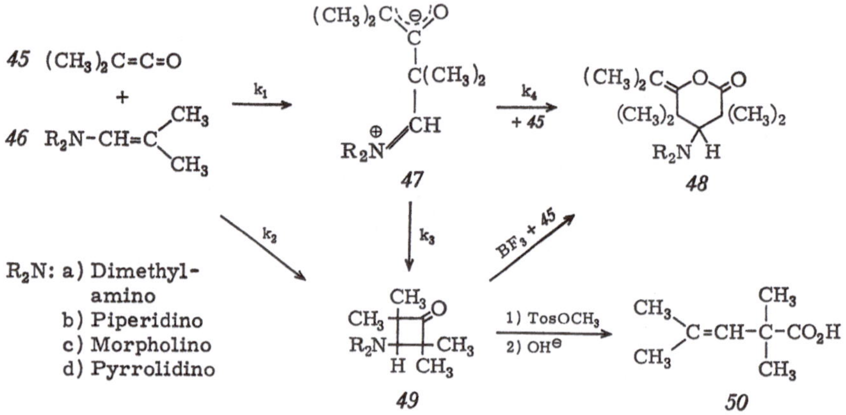

Die *Cyclobutanon-Struktur* von 49 folgt u. a. aus der typischen IR-Bande bei $\approx$ 1770 cm^{-1} sowie der alkalischen Spaltung des Quaternierungsproduktes zur ungesättigten Säure 50. Der Methylen-δ-lacton-Charakter von 48 ist durch eindeutige Reaktionen gesichert [47].

Bereits die Bildung von 48 läßt auf die zwitterionische Zwischenstufe 47 schließen, die entweder unter 1.4-Addition weiterer Dimethylketens zu 48 führt oder unter Ringschluß zu 49.

Tatsächlich läßt sich das Verhältnis 49/48 durch wachsenden Dimethylketen-Überschuß zugunsten von 48 beeinflussen. Der Versuch ist beweiskräftig, da 49 mit überschüssigem Dimethylketen erst auf Zusatz von Bortrifluorid zu 48 reagiert [48].

Aus der quantitativen Auswertung der Lösungsmittelabhängigkeit des Verhältnisses 49:48 folgt, daß ein bestimmter Bruchteil von 49 dem Solvenseffekt nicht unterliegt, obwohl mit steigender Solvenspolarität mehr 48 entsteht, die Lebensdauer von 47 also zunimmt. Das legt den Schluß nahe, daß 49 sowohl in einstufiger Cycloaddition als auch über 47 entsteht [49]. Derartige synchrone 2 + 2-Cycloadditionen sind unter Beteiligung von Ketenen als thermische Reaktionen erlaubt [50]. Ihre Stereospezifität findet sich aber prinzipiell auch bei der Reaktion über zwitterionische Zwischenstufen vom Typ *47*, die infolge von „Homoallyl-Mesomerie stabilisiert sind [51]. Außerdem erlauben im Normalfalle die Orbitalsymmetrien nur den Ringschluß zum Cyclobutanon und nicht zum Oxetan [52]. Diese beiden möglichen Cycloaddukte wurden als analoge Reaktionsprodukte nur im Falle der vergleichbaren Reaktion zwischen *51* und dem Inamin *52* zu *53* und *54* beobachtet [53].

51 $(C_6H_5)_2C=C=O$

52 $(C_2H_5)_2N-C\equiv C-R$ $\rightarrow$

$(C_6H_5)_2C-C=O$
$(C_2H_5)_2N$C=CR
53

$+$

$(C_6H_5)_2C=C-O$
RC=C$-N(C_2H_5)_2$
54

Mit *Diphenylketen* (*51*) als Partner wird sogar der Cyclobutanonringschluß zu *58* rückläufig: In Acetonitril verwandelt sich *58* durch überschüssiges *51* in das 2:1 Addukt *57*.

51 $(C_6H_5)_2C=C=O$

$+$

46 $R_2N-C=C(CH_3)_2$
      H

$R_2N = Morpholino$

$\longrightarrow$

$(C_6H_5)_2C$
55 $R_2\overset{\oplus}{N}=CH$
$C(CH_3)_2$

$+ 51$

$\longrightarrow$

C_6H_5 H O
(CH$_3$)$_2$
56 H NR$_2$

$(C_6H_5)_2$
R_2N(CH_3)_2
H
58

$(C_6H_5)_2C$ O O
$(CH_3)_2$(C$_6$H$_5$)$_2$
H NR$_2$
57

Außerdem geht *58* beim Erwärmen in Methanol, mit Salzsäure schon in der Kälte, unter intramolekularer Substitution in *56* über. In beiden Fällen muß man das Zwitterion *55*, dessen Enolatstruktur hier noch besser als bei *47* (S. 244) stabilisiert ist, als Zwischenstufe verantwortlich machen [48].

Nach dem gleichen Additionsprinzip sind zahlreiche 2.4-persubstituierte Cyclobutanone der allgemeinen Formel *59* dargestellt worden [54].

R^1 O
R^2
)N R^3
H R^4
59

H O
R^2
)N R^3
H R^4
60

H O
H
)N R^3
H R^4
61

H O
R^2
)N R^3
H H
62

Zugleich wurde ein unabhängiger Beweis für die Vierringstruktur der Addukte erbracht, wie das folgende Beispiel zeigt [57]. Die Kombinationen 63 + 64 sowie 66 + 45 führen zum gleichen Produkt 65. Hier, wie in

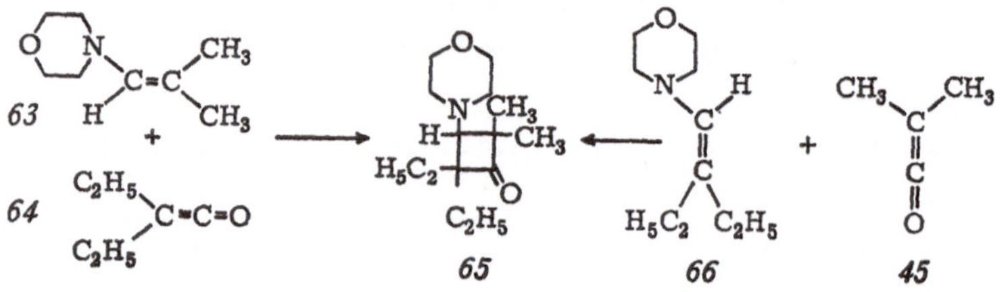

zahlreichen anderen Fällen wurde das Keten in situ aus dem entsprechenden Säurechlorid und Triäthylamin erzeugt. Durch Variation der Komponenten sind auch die Cyclobutanone vom Typ 60 [55], 61 [56—58] und 62 [46,55,56,57] zu fassen. Entscheidend ist dabei die Verwendung der unpolaren Solventien Äther oder Benzol, in denen das aus dem Säurechlorid entstandene Triäthylammoniumchlorid quantitativ ausfällt, für die Synthese der Typen 60—62: Mit steigender Zahl von H-Atomen neben der Carbonylgruppe wächst die Instabilität der Cyclobutanone gegen polare Solventien und Protonenquellen (s. u.) sowie Temperatursteigerung. Es tritt zwischen 50—110° C Ringöffnung ein, die beim Typ 60 und 61 eindeutig verläuft, z. B. 67→68 [46] und 69→70 [57]. Dagegen führt die Spaltung des Typs 62 zu Gemischen von Acylenaminen, z. B. 71→72+73 [57].

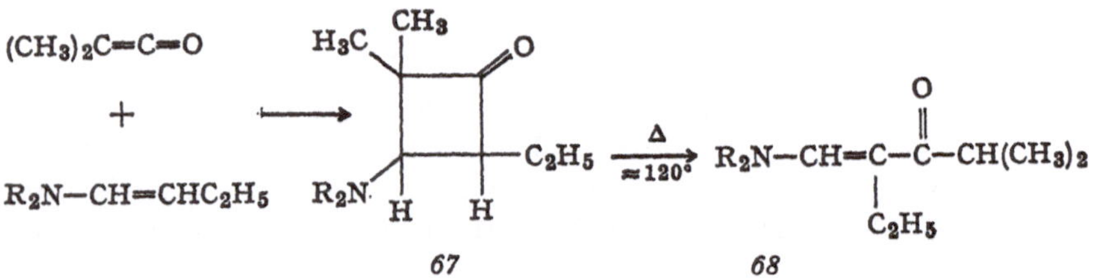

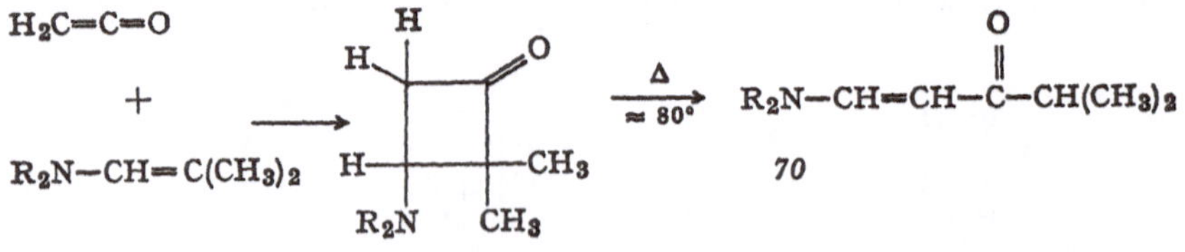

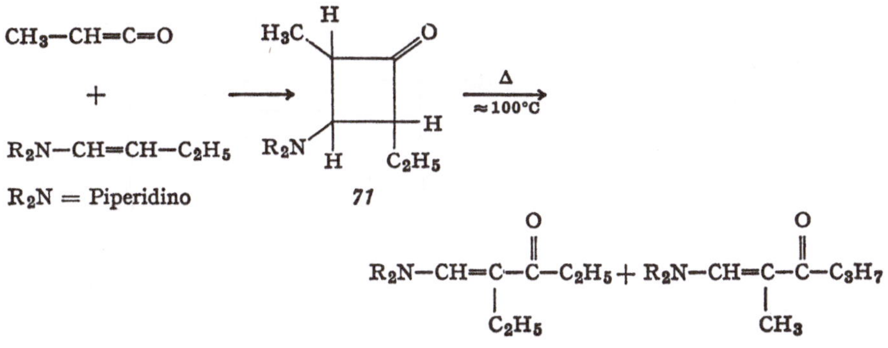

$$CH_3-CH=C=O$$

+

$$R_2N-CH=CH-C_2H_5$$

$R_2N =$ Piperidino

71

$$R_2N-CH=C-\overset{O}{\overset{\|}{C}}-C_2H_5 + R_2N-CH=C-\overset{O}{\overset{\|}{C}}-C_3H_7$$
$$\quad\quad\quad\underset{C_2H_5}{|}\quad\quad\quad\quad\quad\quad\quad\underset{CH_3}{|}$$

72   **73**

Noch weniger substituierte Cycloaddukte wie *74* sind nur noch als kurzlebige Zwischenstufen nachzuweisen (IR); so isomerisiert *74* rasch zu *75* und *76* [46)].

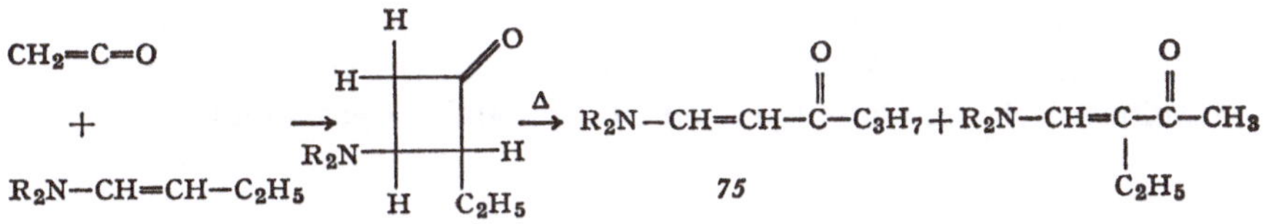

$$CH_2=C=O$$

+

$$R_2N-CH=CH-C_2H_5$$

$R_2N =$ Piperidino

74

$$R_2N-CH=CH-\overset{O}{\overset{\|}{C}}-C_3H_7 + R_2N-CH=C-\overset{O}{\overset{\|}{C}}-CH_3$$
$$\quad\quad\quad\quad\quad\quad\quad\quad\quad\quad\quad\quad\quad\quad\quad\underset{C_2H_5}{|}$$

75   **76**

Die Spaltung der 3-Aminocyclobutanone, die in Nachbarschaft zur Carbonylgruppe über H-Atome verfügen, muß nicht unbedingt über die bei der Bildung durchlaufenen Zwitterionen (vgl. S. 12 *55⇌58*), führen. Das Problem ist an den ähnlich gebauten Cycloaddukten der Ketonenamine untersucht worden (vgl. 2.2).

Die eindeutigen Spaltungen besitzen *präparatives* Interesse für die Synthese von Acylenaminen. Dabei ist zu beachten, daß im Falle der Bildung von *68* sich das Ergebnis nicht von dem einer direkten Acylierung des eingesetzten Enamines unterscheidet. Im Falle von *70* dagegen sind die eingesetzten Komponenten an der olefinischen Bindung gespalten worden, sodaß also Enamin- und Acylfunktion gewissermaßen vertauscht erscheinen. Gerade hier ist dieser Syntheseweg bedeutsam, weil das unsubstituierte Enamin *78* viel zu leicht Selbstkondensationen (vgl. Kap. 1.3, S. 242) unterliegt. Erst wenn man durch Acylsubstituenten die Basizität des Stickstoffatoms schwächt (z.B. *79, 80*) wird diese Ausweichreaktion vermieden und normale „Acylierung" möglich, z.B. *77 + 79→81 →83* und *77 + 80→82* [55a)].

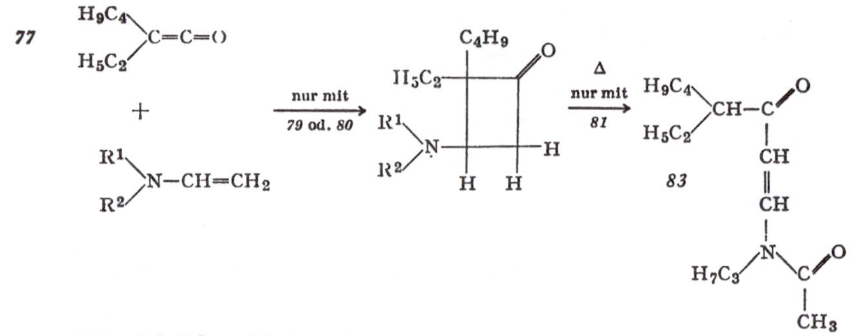

77
$$H_9C_4 \atop H_5C_2 \!\!> C=C=O$$

$+$

$$R^1 \atop R^2 \!\!> N-CH=CH_2$$

$\xrightarrow[\text{79 od. }80]{\text{nur mit}}$

(Cyclobutanone-Ring: C_4H_9, H_5C_2, R^1, R^2, N, O, H, H)

$\xrightarrow[\text{81}]{\Delta \atop \text{nur mit}}$

$$H_9C_4 \atop H_5C_2 \!\!> CH-C \!\!< \!\!O$$
$$|$$
$$CH$$
$$\|$$
$$CH$$
$$|$$
$$H_7C_3 \!\!> N-C \!\!< \!\!O$$
$$|$$
$$CH_3$$

83

78: R^1, R^2 = Alkyl
79: $R^1 = C_3H_7$
   $R^2 = COCH_3$ *81*
80: $R^1 = CH_3$ *82* (85%)
   $R^2 = SO_2\text{-}O$

Auf einem ungewöhnlichen, indirekten Weg ist dennoch die Acylierung von *78* (s. o.!) möglich: 1.3-Bis-dimethylamino-buten-1 (*84*), das aus Crotonaldehyd und Dimethylamin bequem zugänglich ist [7] und das sich als Dimeres des Vinyldimethylamins *85* auffassen läßt (vgl. S. 243), verbindet sich mit Dimethylketen zu dem labilen Cycloaddukt *86*.

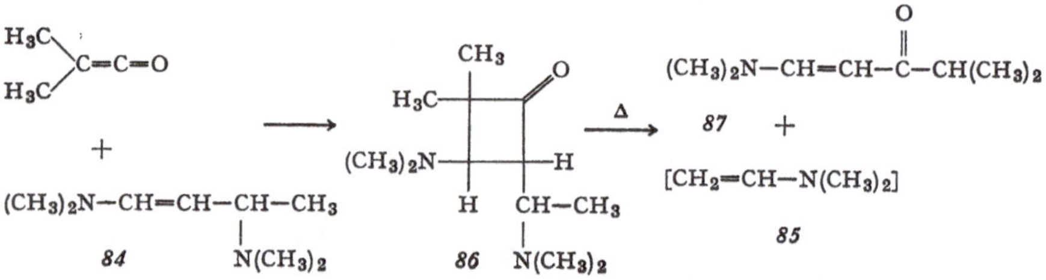

$$H_3C \atop H_3C \!\!> C=C=O$$

$+$

$$(CH_3)_2N-CH=CH-CH-CH_3$$
$$|$$
$$N(CH_3)_2$$

84

$\longrightarrow$

(Cyclobutanone-Ring: CH_3, H_3C, $(CH_3)_2N$, O, H, $CH-CH_3$, H)
$$|$$
$$N(CH_3)_2$$

86

$\xrightarrow{\Delta}$

$$(CH_3)_2N-CH=CH-\overset{\displaystyle O}{\overset{\|}{C}}-CH(CH_3)_2$$

87 $+$

$[CH_2=CH-N(CH_3)_2]$

85

Dieses spaltet thermisch zum Acylenamin *87* und harzigen Produkten, die aus *85* entstanden sein dürften [55a]. Der übliche Weg zu Acylenaminen *88*, die sich formal von *78* ableiten, ist jedoch ganz anders: Man setzt ein Hydroxymethylenketon *89* mit einem sec. Amin um [56b].

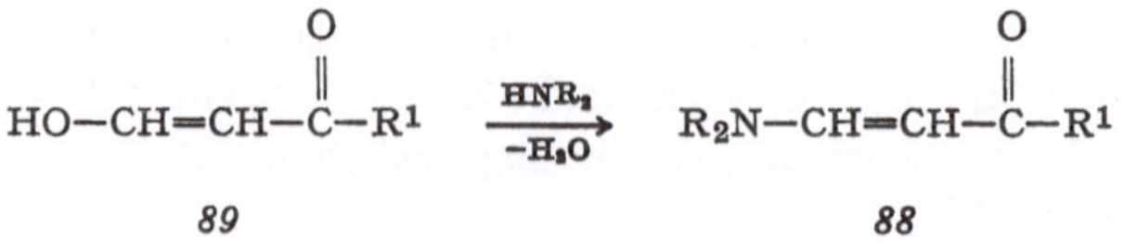

$$HO-CH=CH-\overset{\displaystyle O}{\overset{\|}{C}}-R^1 \xrightarrow[-H_2O]{HNR_2} R_2N-CH=CH-\overset{\displaystyle O}{\overset{\|}{C}}-R^1$$

89   *88*

2.1.2. Acylierung von „Ketonenaminen"

Die Isolierung von Cycloaddukten aus Ketenen und Ketonenaminen ist erst kürzlich gelungen. Während die Reaktion von *90* mit Methylketen zu *91*, das sich zu *92* spalten läßt [59] keine Besonderheit bietet, verdienen die

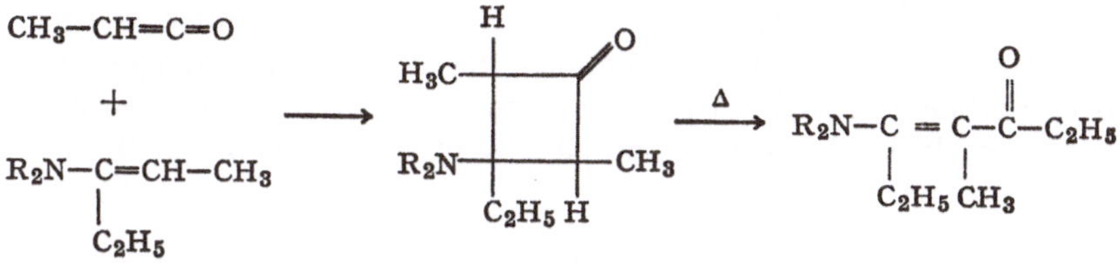
CH3—CH=C=O
+
R2N—C=CH—CH3
C2H5
H
H3C
O
R2N
CH3
C2H5 H
Δ
O
R2N—C = C—C—C2H5
C2H5 CH3

90 *91* *92*

R_2N = Morpholino

Addukte der cyclischen Enamine *93* besondere Beachtung. Auch hier muß die zwitterionische Vorstufe *94* postuliert werden, die entweder unter rascher Protonenverschiebung zu einem Gemisch der Acylenamine gleicher Ringgröße (*95a* und *95b*) weiterreagiert oder zum Cycloaddukt *96*. Dieses spaltet bei $n = 6–8$ ausschließlich zu einer Mischung von *95a* und *95b*) auf.

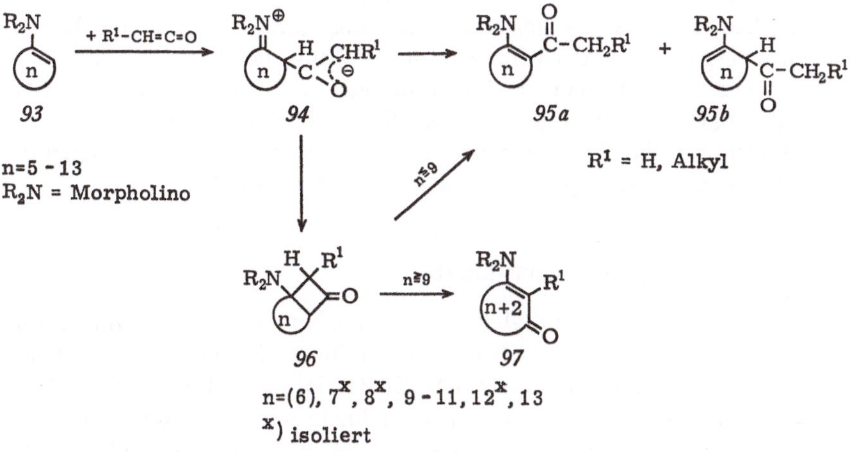
R2N
n
+ R1-CH=C=O
R2N⊕
H CHR1
C ⊖
O
R2N O
C-CH2R1
+
R2N
H
C-CH2R1
O
93
94
95a
95b
n=5 - 13
R2N = Morpholino
n≧9
R1 = H, Alkyl
H R1
R2N
=O
n≧9
R2N
R1
n+2
O
96
97
n=(6), 7x, 8x, 9 - 11, 12x, 13
x) isoliert

Bei einer Ringgröße $n \geqq 10$ erfolgt in Chloroform/Triäthylamin (und Äther/Triäthylamin) die Stabilisierung unter Ringerweiterung zu *97*, während bei $n = 9$ ein Gemisch von *95a, b* und *97* entsteht [60]. In Chloroform/Triäthylamin ist selbst während der Reaktion, auch bei 0 °C, kein

Cycloaddukt nachweisbar [61]. Der Rückschluß auf *96* ist jedoch aus dem Auftreten von *97* bei $n \geqq 9$ zwingend. Erst in Benzol oder Äther ist *96* nachzuweisen und mit $n = 7{,}8$ und 12($R^1 = H$, CH_3) sogar zu isolieren [59]. Die Spaltung der Addukte verläuft in der erwarteten Richtung (vgl. 2.2). Die im Reaktionsgemisch von *93* ($n = 6$) und Keten beobachtete Bande bei ~ 1775 cm^{-1} darf vermutlich dem sehr instabilen Cycloaddukt *96* ($n = 6$, $R^1 = H$) zugeschrieben werden [59,62]. Dagegen dürfte der Ringschluß des Zwitterions für $n = 5$ infolge zu großer Ringspannung ausbleiben. Aus einem ähnlichen Grunde entsteht aus der Fischerbase und Keten oder Acetylchlorid nur das „normale" Acylierungsprodukt *98* [1].

98

Eine größere Stabilisierung des Zwitterions *94* erhöht erwartungsgemäß den Anteil an *95a* + *b* auf Kosten von *97* wie die Ergebnisse an Cyclodecanonenaminen zeigen [19]:

1. Der stärker basische Pyrolidinrest, an Stelle des Morphodinrestes, steigert das Verhältnis *95* : *97* von $\approx 2 : 1$ auf $\approx 20 : 1$ ($R^1 = H$).

2. Die gleiche Steigerung beobachtet man für *93* ($n = 12$) und Keten beim Übergang von Äther auf Acetonitril/Äther (4 : 1).

3. Setzt man Ketene (in Substanz oder in situ) ein mit $R^1 = CH(CH_3)_2$, C_6H_5, p-$CH_3OC_6H_4$ sowie Dimethyl- und Diphenylketen, so entstehen nur die *95a,b* entsprechenden „normalen" Acylierungsprodukte. Intermediäre Cycloaddukte sind nicht nachweisbar.

2.2. Zur Spaltung der Cycloaddukte

Da die Spaltungsgeschwindigkeit der Cyclobutanone mit wachsender Zahl von H-Atomen neben der CO-Gruppe zunimmt, ist damit zu rechnen, daß die Ringöffnung nicht nur über Zwitterionen (z. B. *96* → *94* (S. 249) als Umkehrung der Bildung verläuft. Vielmehr wäre es denkbar, daß als geschwindigkeitsbestimmender Schritt eine Enolisierung erfolgt, z. B. *96* → *96a* oder *96b*, der sich eine elektrocyclische Ringöffnung zu *99* oder *100* anschließt, die infolge der anwesenden Substituenten durchaus wesentlich rascher als die übliche thermische Cyclobuten-Ringöffnung [63] verlaufen könnte. Dabei müßte die Enolisierung *96* → *96a* (oder *96b*) geschwindigkeitsbestimmend sein, da ein H-D-Austausch in *96* nicht gelingt. *96* ($n = 8$, $R^1 = H$) reagiert jedoch nur 1,5mal rascher zu *95* ($n = 8$, $R^1 = H$) als das deuterierte Analoge *101* ($n = 8$, $R^1 = H$) [59]. Für den geschwindigkeitsbestimmenden Enolisierungsschritt *96* → *96 a (b)* wäre ein viel größerer Isotopieeffekt zu erwarten. Daher ist der Weg über *96 a (b)* auszuschließen. Das Ergebnis steht mit einem sekundären Isotopeneffekt für einen Übergangszustand im Einklang, der der zwitterionischen Zwischenstufe *94* nahe steht. Die Spaltungsgeschwindigkeit hängt stark von der Ringgröße *n* ab, wie Tabelle 1 zeigt.

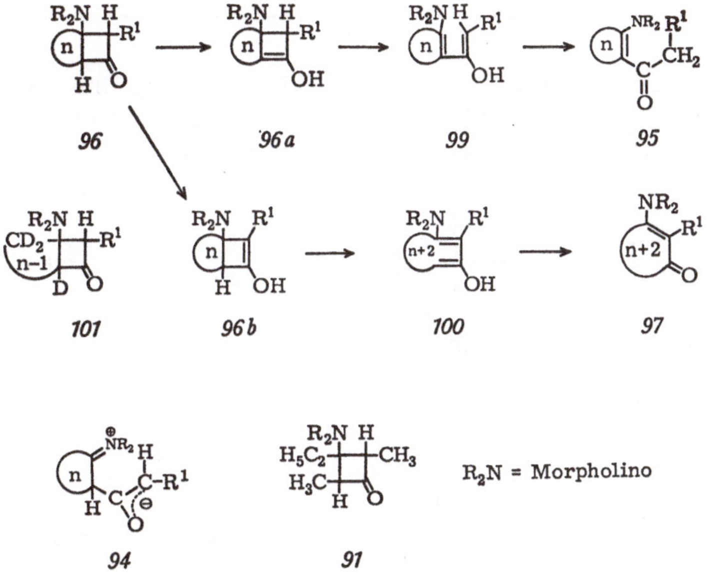

Tabelle 1. *Halbwertszeiten verschiedener Cycloaddukte in Tetrachloräthylen bei 20 °C*)*

*96***) ($n = 7$)	*96***) ($n = 8$)	*96***) ($n = 12$)	*91*
> 5 Tage	115 min	230 min	1400 min

*) Reaktion 1. Ordnung
**) $R^1 = CH_3$

Sieht man von der ungewöhnlichen Stabilität von *96* ($n = 7$) ab, so öffnen sich die bicyclischen Verbindungen rascher, als die Modellsubstanz *91* [59].

Das Solvens beeinflußt, wie am Beispiel des Chloroforms gezeigt, die Spaltungsgeschwindigkeit erheblich, die außerdem durch Protonenkatalyse nochmals erhöht wird (Tabelle 2).

Tabelle 2. *Halbwertszeiten der Isomerisierung von 96 (n = 7, $R^1 = H$) zu 95 in Tetrachloräthylen + Zusätze bei 20 °C (Reaktion 1. Ordnung)*

Zusatz	—	10 Moläquiv. Methanol	1 Moläquiv. Phenol	1 Moläquiv. 3.5-Dichlorphenol
t $^1/_2$	> 8 Tage	~ 500 min	~ 150 min	~ 30 min

Sehr eigentümlich ist der Solvenseffekt auf die Ringöffnungsrichtung der Cycloaddukte *102* und *103*, der sich durch quantitative Bestimmung des auf dem Wege *102* (oder *103*) → *105* → *106* gebildeten Cyclododecanons

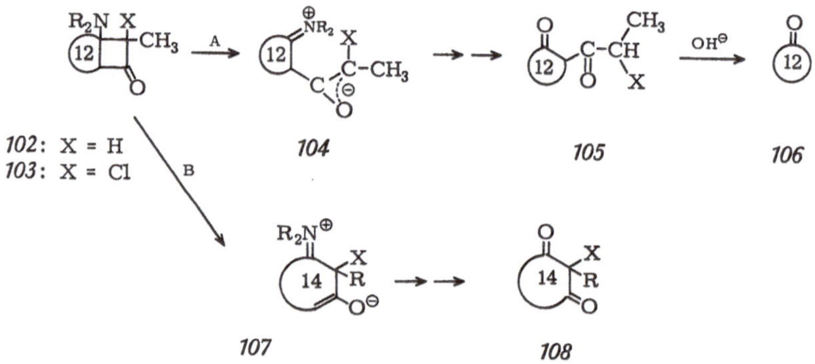

102: X = H
103: X = Cl

erfassen läßt. Wie Tabelle 3 lehrt, nimmt der Anteil des Weges A für *102* mit steigendem E_T [64)]-Wert ab, während für *103* das Umgekehrte gilt. Die Sonderstellung des Chloroforms, wahrscheinlich auf Spuren Chlorwasserstoff beruhend, verschwindet auf Zusatz von wenig Triäthylamin, also unter den meist für präparative Zwecke benutzten Bedingungen (vgl. *93* bis *97*, S. 249).

2.3. Eigenschaften und Reaktionen der Acylenamine

2.3.1. Physikalische Daten der Acylenamine

Spektroskopische Untersuchungen an zahlreichen sterisch weitgehend fixierten Acylenaminen zeigen, daß sich die möglichen planaren Anordnungen teilweise so deutlich in der Lage der IR- und UV-Banden

Tabelle 3. *Ausbeute (%) an Cyclododecanon aus 102 bzw. 103 in Abhängigkeit vom Solvens*

Solvens	102	103	E_T
Diäthyläther	35	9	34,6
Essigsäureäthylester	9	9	38,1
Pyridin	7	13	40,2
Acetonitril	7	26	46,0
Äthanol	4	23	51,9
Chloroform	23	—	39,1
Chloroform + Triäthylamin 30:1	7	—	(39)

wiederspiegeln [65a,66]), daß sie Rückschlüsse auf die bevorzugten Konfigurationen und Konformationen beweglicher Acylenamine erlauben.

trans-s-trans   cis-s-trans   trans-s-cis   cis-s-cis

So findet sich in den *IR-Spektren* die CO-Bande für die trans-s-trans-Anordnung (*109* [65]), *110* [65a,65b]) bei 1600—1630 cm^{-1}, für den cis-s-trans-Typ bei ~1640 cm^{-1} (*111*, *111 a* [66a]) und schließlich für die trans-s-cis Anordnung bei ~1655 cm^{-1} (*112* [66]). Die längstwelligen CO-Banden (1580—1600 cm^{-1}: *261* [122])) werden beim cis-s-cis Typ beobachtet. Allerdings sind Acylenamine diesen Typs mit eindeutiger Lage der Substituenten nur bekannt, wenn die sec. Aminogruppe zur CO-Gruppe eine intramolekulare H-Brücke ausbilden kann.

Die Lage der Bande, welche der konjugierten C=C-Bindung zugeschrieben wird, ist dagegen weit weniger charakteristisch (1545—1580 cm^{-1}).

Im *UV-Spektrum* setzt sich lediglich die trans-s-trans-Form deutlich von den übrigen ab, indem sie ~30 nm kürzerwellig bei deutlich höherem ε absorbiert (*109, 110* [65,65a,65b]) (vgl. jedoch [65a]). Die Wirkung unterschiedlicher Aminreste (z.B. Pyrrolidin-Morpholin) muß außerdem berücksichtigt werden.

Für das offenkettige Enaminoketon *88 a*, bei dem auf Grund des NMR-Spektrums sicher ist, daß Aminogruppe und Acylrest transständig

	109 [65)]	110 [65a, 65b)]	111 [66a)]
	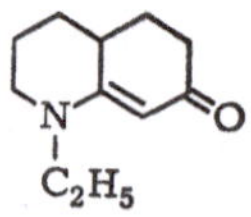 C_2H_5	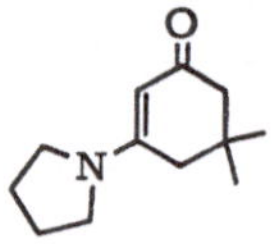	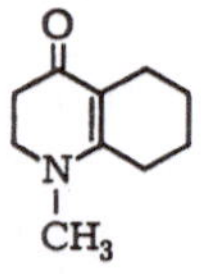 CH_3
Konfiguration Konformation	trans-s-trans	trans-s-trans	cis-s-trans
UV: λ_{max} (nm), (log ε) Lösungsmittel	304 (4.97) Äthanol	303 (4,54) Äthanol	336 (4.16) Äthanol
IR: $\tilde{\nu}$ (cm^{-1}) Lösungsmittel	1600 (C=O) 1550 (C=C) CH_2Cl_2	1607 (C=O) 1560 (C=C) Nujol	1640 (C=O) 1580 (C=C) —

	111a [66a)]	112 [66)]	261 [122)]
	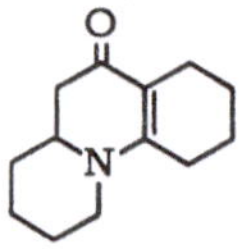	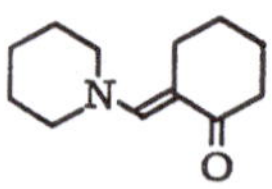	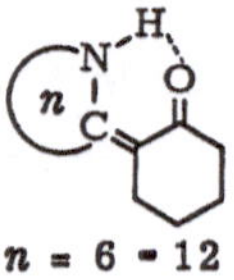 $n = 6 - 12$
Konfiguration Konformation	cis-s-trans	trans-s-cis	cis-s-cis
UV: λ_{max} (nm), (log ε) Lösungsmittel	336 (4.15) Äthanol	335 (4.23) Äthanol	330—338 (4.28—431) Methanol
IR: $\tilde{\nu}$ (cm^{-1}) Lösungsmittel	1640 (C=O) 1544 (C=C) CCl_4	1654 (C=O) 1545 (C=C) C_2Cl_4	1580—1600 (C=O) 1560—1570 (C=C) CH_2Cl_2

sind (Kopplungskonstante der Vinyl-Wasserstoffe: $J = 12{,}5$ Hz), findet man im IR-Spektrum der kristallinen Verbindung die Banden $\nu_{C=O} = 1631$ cm^{-1} und $\nu_{C=C} = 1584$ cm^{-1}, welche charakteristisch für die trans-s-trans-Form sind.

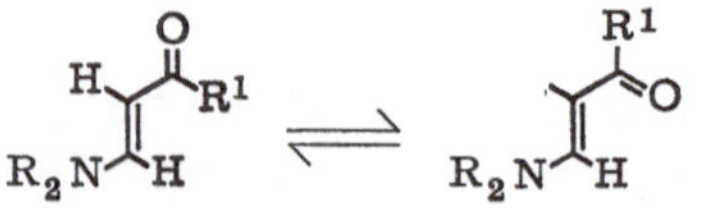

1631 (C=O) 1584 (C=C) trans-s-trans	1654 (C=O) 1586 (C=C) trans-s-cis

88a $R = R^1 = CH_3$
88b $R = CH_3$, $R^1 = C_2H_5$

Acylenamin	λ_{max}(nm, ε) Lös. Mittel	IR (ν_{max})	Zuordnung
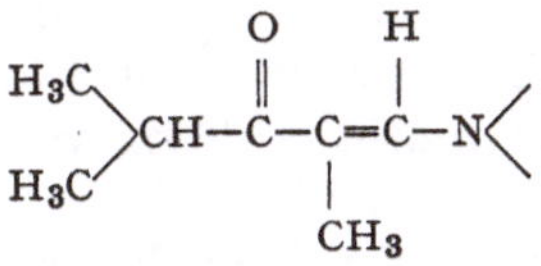 **113** $H_3C{-}$ / $H_3C{-}$ CH–C(=O)–C(CH$_3$)=C(H)–N<	287(18300) Hexan Morpholin 291(19800) Hexan Piperidin	1625, 1600—1580 (Film)	trans-s-trans
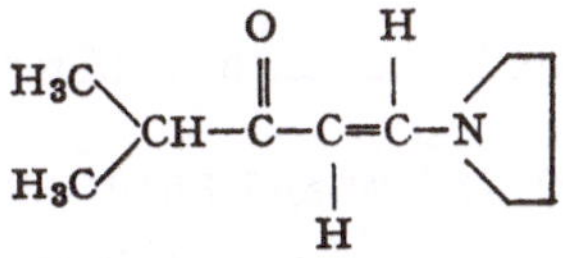 **114** $H_3C{-}$ / $H_3C{-}$ CH–C(=O)–C(H)=C(H)–N (pyrrolidine)	295(28100) Hexan	1650, 1610, 1570 in KBr	trans-s-cis und trans-s-trans
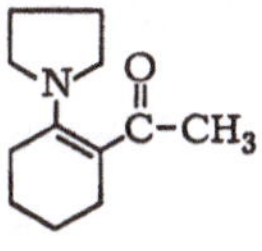 **115** (pyrrolidine enamine of cyclohexanone, C(=O)–CH$_3$)	316	1610, 1492	cis-s-trans s. G. Opitz [68] als kristalline Substanz
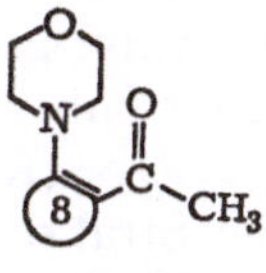 **116** [59] (morpholine enamine, 8-membered ring, C(=O)–CH$_3$)	362 (8250) Methanol 218 (6900)	1590, 1525 in KBr	cis-s-trans als kristalline Verbindung
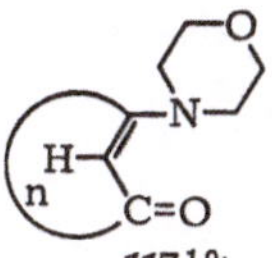 **117** [19] (morpholine, ring size n, H–C=, C=O) n = 12, n = 13, n = 14	308 (17900) Dioxan 308 (20800) Dioxan 307 (21500) Dioxan	1655 s, 1540 s in C$_2$Cl$_4$	cis-s-cis
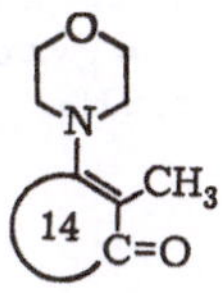 **118** [19] [59] (morpholine, ring 14, CH$_3$, C=O)	317 (4860) 209 (4660) Schulter bei 225 nm	1685 s, 1630 s in KBr	Spektrum eines vinylogen Säureamids mit verdrillten Substituenten

Welches Konformere bei Enaminoketonen des Typs *88* im kristallinen Zustand bevorzugt ist, die s-cis oder die s-trans-Form, ist abhängig von der Größe des Substituenten R^1. Geht man von $R^1 = CH_3$ zu $R^1 = C_2H_5$ (*88b*) über, so tritt im IR-Spektrum des kristallinen Produktes die $\nu_{C=O}$-Bande bei 1654 cm^{-1} und die $\nu_{C=C}$-Bande bei 1586 cm^{-1} auf. Das bedeutet, daß im kristallinen Zustand *88b* in der trans-s-cis-Form vorliegt.

In den Lösungen von *88a* und *88b* liegen Konformen-Gemische vor, kenntlich am Auftreten aller drei Banden ($\nu_{C=O}$ s-cis, $\nu_{C=O}$ s-trans und $\nu_{C=C}$).

Aus diesen Ergebnissen lassen sich die Konformationen weiterer Acylenamine ableiten.

Die Zuordnung für die Verbindungen *115* [68], *116* [59] und *118* [19,61] läßt sich nur für den kristallinen Zustand treffen, da in Lösung noch ein Ausweichen der Doppelbindung in die unkonjugierte Position möglich ist. *118* zeigt, wie ein 2-ständiger Substituent im Vergleich zu *117* eine Verdrillung der Substituenten herbeiführen kann, wenn die Enden des

$>C=C-C=O$-Systems durch einen Ring festgehalten werden. Auch hier

gestattet das IR-Spektrum eine Aussage zur Geometrie des Enaminoketons *118*.

Eine Sonderstellung nehmen die Acylenamine von Typ *95* (S. 249) ein. Die erzwungene cis-Anordnung der Substituenten verhindert infolge mangelhafter Planarität den vollen Energiegewinn der vinylogen Carbonamidstruktur, so daß die Doppelbindung teilweise in die unkonjugierte Stellung ausweicht. So zeigt das Isomerenpaar $95\,a_A \rightleftarrows 95\,a_B$ ($R^1 = H$) unabhängig von der Ringgröße die typische Bande der Struktur $95\,a_A$ (358 nm, $\varepsilon \approx 8000$ [CH_3OH]) nur mit geringer Intensität, während eine recht intensive Enaminbande von $95\,a_B$ auftaucht (217 nm, $\varepsilon \approx 7000$). Dem entspricht eine *normale* Carbonylschwingung bei 1705 cm^{-1} und eine Olefinbande bei ≈ 1641 bzw. 1645 cm^{-1} (z. B. $n = 6$ [68], $n = 7$ [59,67], $n = 8$ [59]).

Aus dem Signal des Vinylprotons im NMR-Spektrum läßt sich auf etwa 90% $95\,a_B$ (R^1=H, n=6) im Tautomerengemisch schließen [68]. Die gleichen Indizien bestätigen die Sonderrolle des Pyrrolidinorestes, der das Gleichgewicht zu 85—90% nach $95\,c_A$ (n=6, R^1=H) treibt [66]. Aber auch der Übergang zum Fünfring ist wirkungsvoll: Selbst mit der Morpholinogruppe liegt nunmehr nur das konjugierte Isomere $95\,a_A$ (R^1=H, n=5) vor [67]. Die makrocyclischen Acylenamine 97 (R^1=H) bevorzugen ausschließlich die gezeigte konjugierte Form, dagegen erzwingt, für n=13, die Einführung einer Methylgruppe vorwiegend die unkonjugierte Form 119 [19].

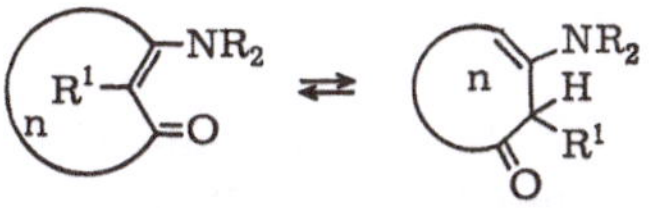

$$R_2N = Morpholino$$
$$R^1 = H \quad n = 12, 13, 14$$
$$R^1 = CH \quad n = 13, 14$$

97   119

2.3.2. Acylierung der Acylenamine mit Ketenen

Sofern die Darstellung der Acylenamine durch Reaktion mit Ketenen erfolgt, muß mit weiterer Acylierung des Acylenamins gerechnet werden, eine Reaktion, die mit überschüssigen Ketenen in den Vordergrund tritt.

Nicht nur die eingangs erwähnten dipolaren Zwischenstufen reagieren mit Ketenen unter 1.4-Addition (vgl. 2.1.1); auch von Acylketonenaminen ist dieser Reaktionstyp bekannt. So scheidet die benzolische Lösung des Enamins 120 beim Einleiten von überschüssigem Keten das Addukt 122 ab, das beim Stehen in das α-Pyron 123 übergeht [58,69]. Als Zwischenstufe ist 121 anzunehmen, da in Substanz eingesetzte Acylenamine in

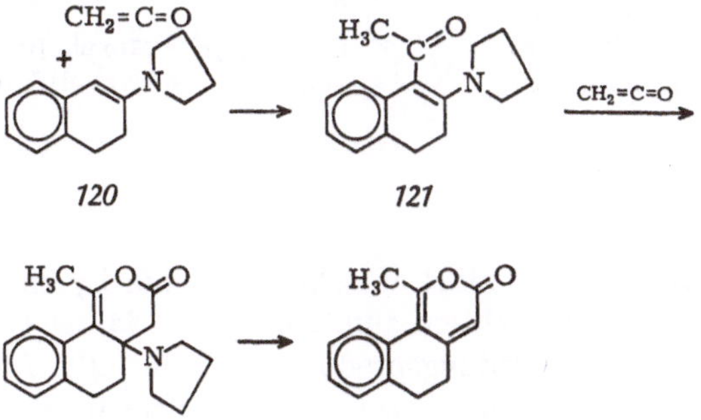

120   121

122   123

gleicher Weise reagieren können (s. u). Die Reaktion fügt sich in das allgemeine Schema der 1.4-Cycloaddition von Ketenen an α,β-ungesättigte Ketone [70,71]. Von zahlreichen Beispielen [56,57,62,69] seien die Übergänge *124*→*125* [69], *97* (R^1 = H, *n* = 14)→*126* [19] und *127*→*129* [69] herausgegriffen.

124, *125*, *97* (R_1=H, n=14), *126*

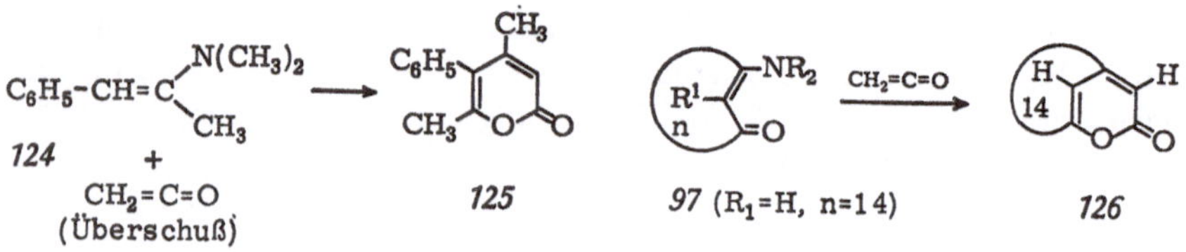

Der letzte Fall ist ungewöhnlich [72], da zunächst eine C-Acylierung zu *128* angenommen werden muß, die nur für $R^1 = OC_2H_5$ zu erwarten ist. Acylenamine, die in das nichtkonjugierte Isomere ausweichen können,

127, *128*, *129*

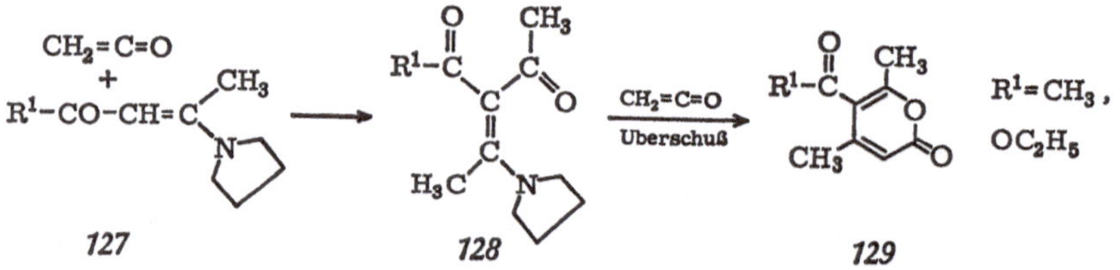

bilden zusätzliche Produkte. Zwar reagiert *130* unabhängig vom Aminrest zum α-Pyron *131*. Im Falle der Acylenamine *133a, b*, hängt jedoch — entsprechend dem Gleichgewicht — das Ergebnis stark vom Aminrest

130, *131*

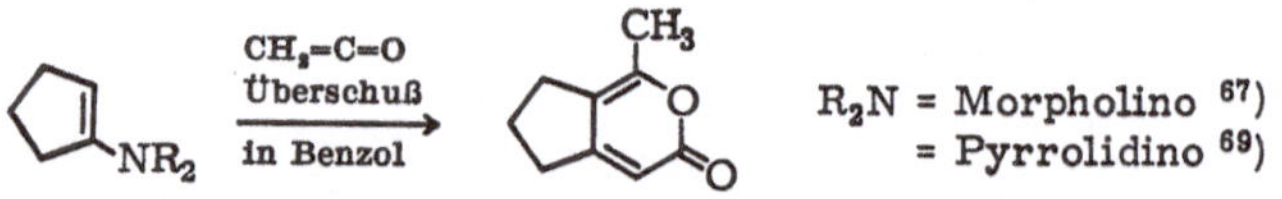

ab [68]. Der Pyrrolidinrest führt ausschließlich zum Pyron *132*, während der Morpholinrest die Reaktion zum Enolester *134* dirigiert [67,68], der sich besser als sein Verseifungsprodukt isolieren läßt. *135* lagert sich sofort zu *136* um, eine Umacylierung [73], die nicht beim 5- und 7-Ring eintritt [67].

The reaction scheme shows structures **132**, **133a**, **133b**, **134**, **137**, **138**, **135**, and **136** with reagents CH$_2$=C=O / Äther, CH$_2$=SO$_2$, and H$_3$O$^\oplus$.

R$_2$N:

50 %	a: Pyrrolidino	38 %	8
7 %	b: Piperidino	9	57
0 %	c: Morpholino	5	66

Diese Umlagerung ist zu berücksichtigen, wenn der Angriff an der Carbonylfunktion des Enolesters dem gezielten Ringschluß zu Pyrazolderivaten dienen soll [67,79]. Auch das Medium kann entscheidend eingreifen. Erzeugt man das Keten in Chloroform mit Triäthylamin, so wird auch 1-Morpholino-cyclopenten glatt in den Enolester verwandelt [75].

Diese Reaktionsführung hat sich auch bei anderen Ringgrößen bewährt [30,73,75].

Wie man sieht, eignet sich Keten nicht zum Abfangen isomerer Acylenamine aus ihrem Gemisch. Für diesen Zweck läßt sich sehr gut die Fähigkeit des aus Methansulfochlorid mit Triäthylamin erzeugten *Sulfens* ausnutzen, sowohl mit Enaminen [76] als auch mit Acylenaminen [68, 76] *stabile* Cycloaddukte zu bilden, wie die Isolierung von *137* und *138* zeigt. Entsprechend den spektroskopischen Befunden läßt sich aus *118* ($n = 14$) mit Sulfen das δ-Sulton *139* isolieren, während das niedere Homologe *97* (R^1 = CH$_3$, $n = 13$, NR$_2$ = Morpholin) das cyclische Sulfon *140* liefert [19].

Structures **139** and **140** with R^{2}N = Morpholino.

259

Unerwartet verhält sich das cis-fixierte Acyl-aldo-enamin *112* ($n=6$) gegenüber Keten. An Stelle der 1.4-Cycloaddition tritt offenbar eine 1.2-Addition zu *141* ein, da *142* als Endprodukt gefaßt wird [77]. Sulfen

143 *112* (n=6) *141* *142*

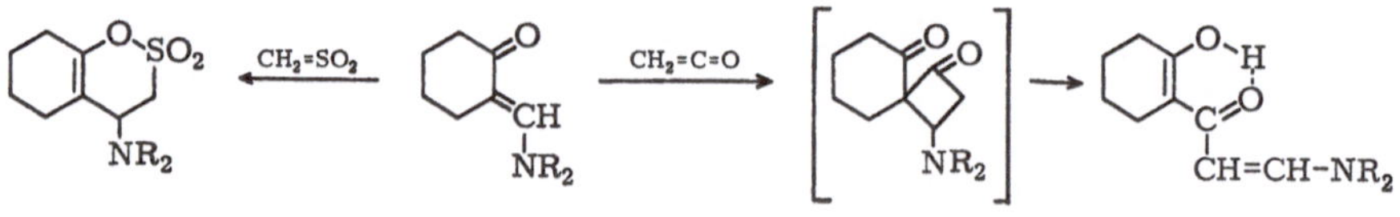

addiert sich dagegen in erwarteter Weise zu *143* [77]. Daß überschüssiges Keten unter forcierten Bedingungen Enamine grundsätzlich zweifach acylieren kann, zeigt die Bildung von 2.6-Dipropionyl-cyclohexanon (*144*) aus *93* ($n=6$) und überschüssigem, in situ erzeugten Methylketen. Es ist anzunehmen, daß der zunächst gebildete Enolester *145* zu *146* weiteracyliert wird [78].

93 (n=6) *145* *146* *144*

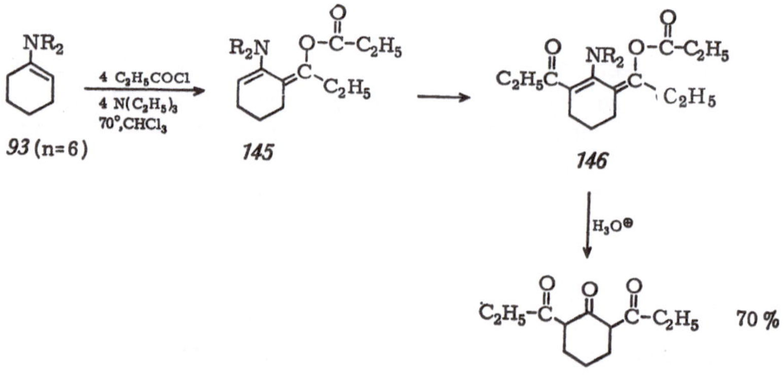

2.3.3. Präparative Bedeutung der Acylenamine

Die beschriebene hydrolytische Abspaltung der Aminogruppe aus den Acylaminoenolestern im sauren Medium läßt sich ebenso glatt auf Acylenamine anwenden. Diese Reaktion ist, wie der Übergang *147* *) → *148* zeigt, eine wichtige Methode zur Gewinnung von *2-Acylcyclanonen* [30,73,73a,75,79,80,81].

*) Die gebräuchliche, wenn auch inkonsequente Schreibweise soll das Gemisch aus konjugiertem und unkonjugiertem Acylenamin symbolysieren.

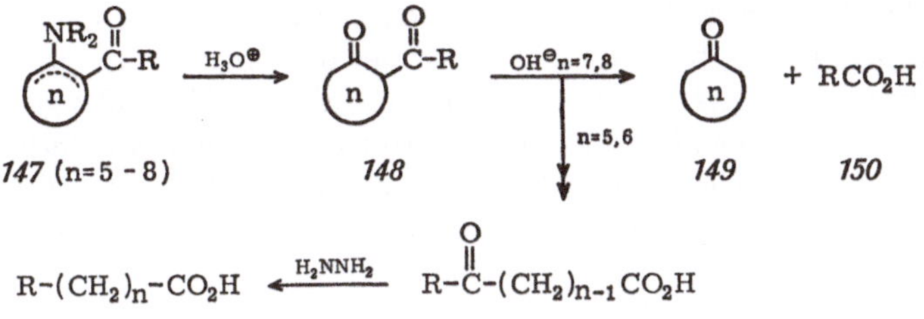

$$147 \ (n=5-8) \qquad 148 \qquad 149 \qquad 150$$

$$152 \qquad\qquad 151$$

Die Spaltung von *148* mit starker Natronlauge [82] erzeugt für $n = 5$ und 6 vorwiegend die Ketosäuren, die sich glatt zur Carbonsäure *152* reduzieren lassen. Damit liegt eine ergiebige Methode vor, um Carbonsäuren, auch verzweigte [80,83] und ungesättigte [81] um fünf [75] oder sechs C-Atome [79] zu verlängern [84].

Die Gesamtausbeuten, bezogen auf eingesetztes Carbonsäurechlorid, betragen 50—60% bei der Kettenverlängerung von Monocarbonsäuren um fünf C-Atome und 45% bei der Kettenverlängerung um sechs C-Atome.

Diese Methode versagt für $n = 7$ oder 8, da das Diketon *148* völlig in die Ausgangskomponenten *149* und *150* gespalten wird [82,85]. Dagegen wird, aus den oben erläuterten Gründen (S. 249) die Methode für Enamine aus Cyclanonen mit $n \geqq 10$ wieder brauchbar, wie das folgende Beispiel zeigt [59,86]. Da die Spaltung *153* → *154* hier ohne Nebenreaktion verlaufen muß, lassen sich mit dem bequem zugänglichen 1-Morpholino-cyclododecen-1 (*93*, $n = 12$) Carbonsäuren mit 70—77% Gesamtausbeute um 12 C-Atome verlängern [86].

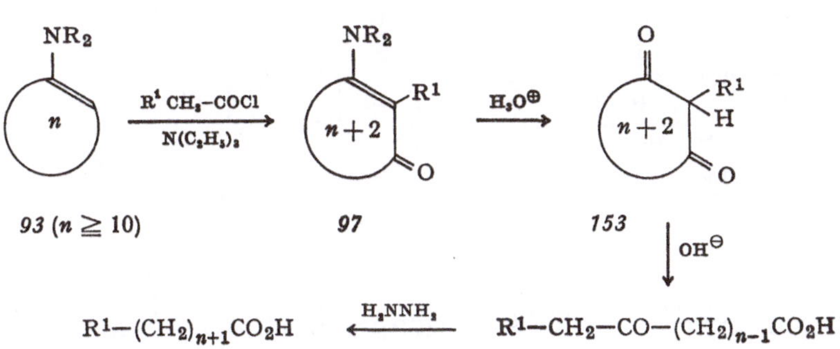

$$93 \ (n \geqq 10) \qquad 97 \qquad\qquad 153$$

$$155 \qquad\qquad 154$$

Alle diese Reaktionen sind grundsätzlich auch auf *Dicarbonsäure-dichloride* übertragbar, von denen hier nur diejenigen zu besprechen sind, die über intermediäre Ketene reagieren. Bei einer Kettenlänge von mindestens 8 C-Atomen (geprüft bis C_{32} [87)]) treten keine Besonderheiten auf. Aus *93* ($n = 5,6$) entstehen über *156* die gut isolierbaren Tetraketone *157* [73,75)], die zu *158* und *159* weiterverarbeitet werden.

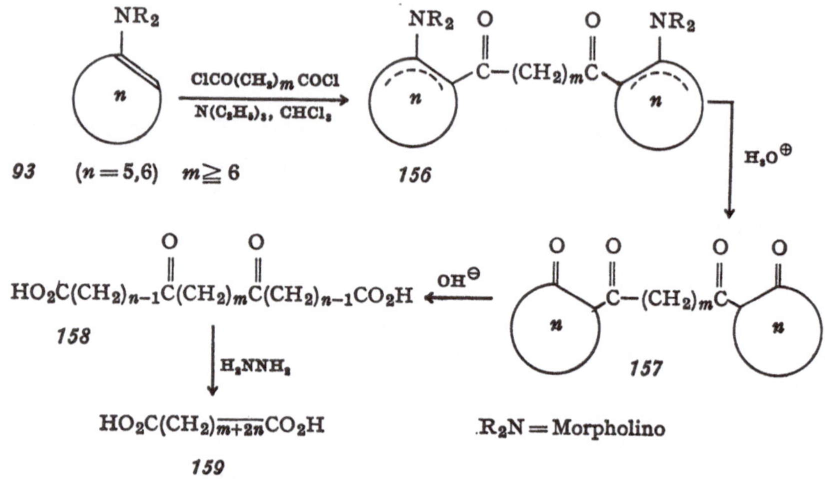

Damit läßt sich die Kette von Dicarbonsäuren mit 40% um 10 und mit 35—45% um 12 C-Atome verlängern. Mit den makrocyclischen Enaminen — geprüft für *93* ($n = 12$) — tritt wieder Ringerweiterung zu *160* ein. Die isolierten Tetraketone *161* werden zu *162* gespalten und schließlich zu *163* reduziert [87)].

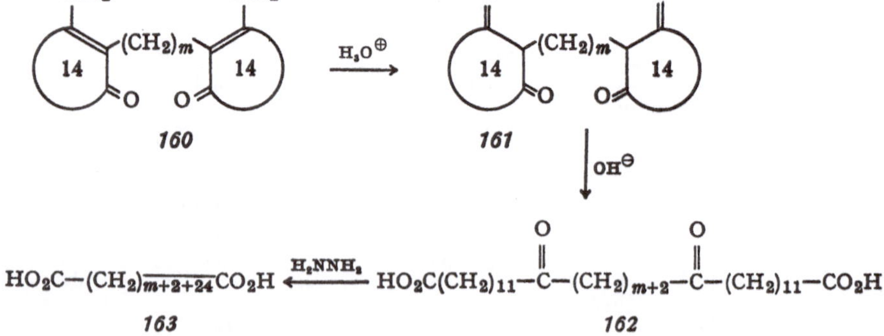

Die genannten Dicarbonsäuren sind auf diesem Wege mit 50% Ausbeute um 24 C-Atome zu verlängern [87].

Bei Verringerung der Kettenlänge der Dicarbonsäuren auf 6 oder weniger C-Atome ändert sich der Reaktionsverlauf. Das zweite aktive Ende greift das gebildete Acylenamin unter Enolesterbildung (164) an [73,75].

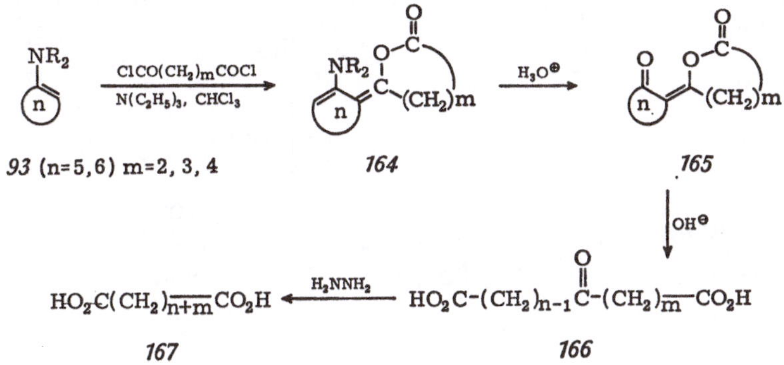

$$93 \ (n=5,6) \ m=2,3,4 \qquad\qquad 164 \qquad\qquad\qquad 165$$

$$167 \qquad\qquad\qquad\qquad 166$$

Die isolierbaren Enolesterketone 165 werden über 166 in 167 umgewandelt. Damit sind Dicarbonsäuren ($m = 3$ und 4) mit $\sim 40\%$ um fünf [75] und mit $\sim 34\%$ um sechs C-Atome [73] zu verlängern. An Stelle des sehr empfindlichen Succindichlorids setzt man besser Bernsteinsäureesterchlorid ein (Gesamtausbeute 42% [75]).

2.4. Acylierung mit Diketenen

Mit Diketen läßt sich der Acetoacetyl-Rest in zahlreiche Nucleophile einführen [88]. Auch von der Fischer-Base ist diese Reaktion schon lange bekannt [1], die zu 168 führt.

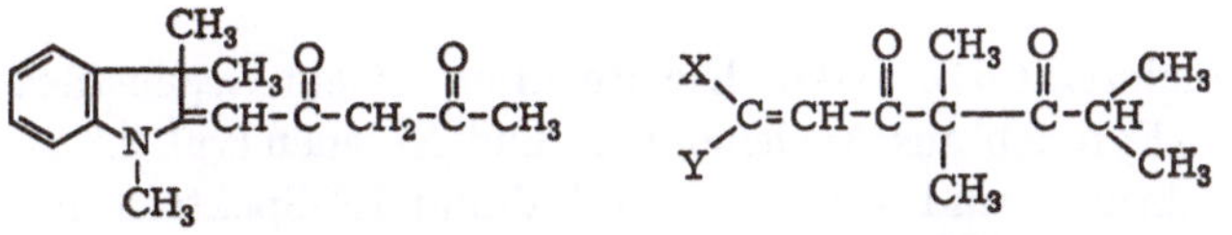

$$168 \qquad\qquad\qquad\qquad 169$$

$$X=C_6H_5, \ OC_2H_5, \ -N\overset{}{\bigcirc}O \quad Y=N\overset{}{\bigcirc}O, \ N(CH_3)_2$$

263

Das β-Lactondimere des Dimethylketens *170* greift enaminartige Verbindungen mit endständiger Doppelbindung in analoger Weise an, wobei z. B. *169* entsteht [89]. Ketonenamine, wie *93* ($n = 5,6$) bilden jedoch mit *170* Cycloaddukte *171* [89], die erst beim Erhitzen Amin zu *172* ab-

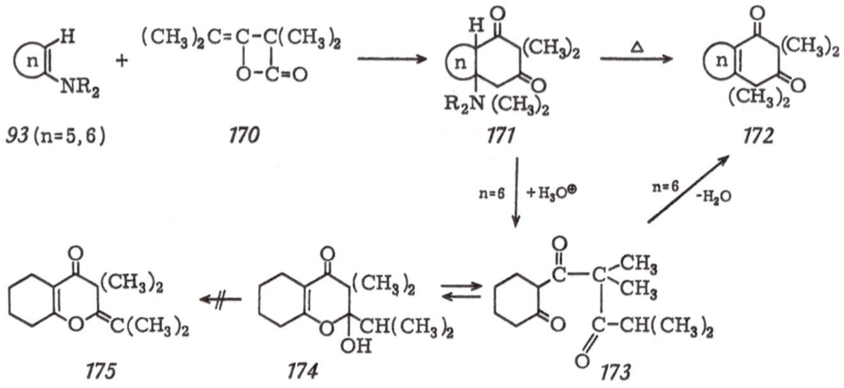

93 (n=5,6) *170* *171* *172*

175 *174* *173*

spalten. Säuren öffnen den Ring von *171* zu *173*, das überwiegend als Halbketal *174* vorliegt. Unter Wasserabspaltung geht es in *172* über und nicht in das Isomere *175* [89]. Mit Enaminen, die kein β-H-Atom besitzen, reagiert *170* nicht, in deutlichem Unterschied zum monomeren Dimethylketen (vgl. 2.1.1.). Beim Einsatz von Diketen *176* selbst ändert sich das Reaktionsbild nochmals. Die hypothetische Zwischenstufe *177* — ein verkapptes 1.3.5.-Triketon — spaltet sofort Amin ab und schließt den

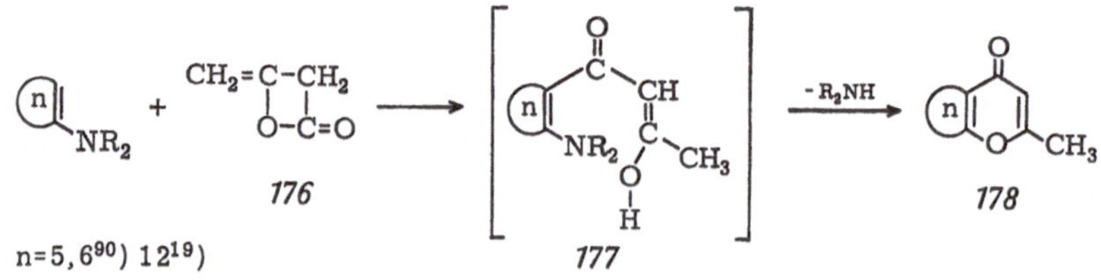

176 *177* *178*

n=5,6[90]) 12[19])

Ring zum γ-Pyronderivat *178* [90,19]. Die Reaktion ist ein Gegenstück zur Bildung von α-Pyronen aus Acylenaminen und Ketenen (vgl. 2.3.2). Die isomeren Strukturen sind anhand ihrer UV-und IR-Spektren eindeutig zu unterscheiden [91,92]. Eine Prüfung ist in jedem Fall erforderlich, da Ketene, besonders wenn sie aus Carbonsäurechloriden und Triäthylamin in situ erzeugt werden, unter Wirkung der Base u. U. schneller dimerisieren [93] als acylieren und Diketene auch Acylenamine angreifen können. Wie kompliziert die Verhältnisse liegen können, zeigt die Umsetzung des Isomerengemisches *118a* und *118b* mit Diketen (*176*) und Methyldiketen (*179*): Verbindung *176* greift das konjugierte Isomere *118a* unter O-Acylierung an und bildet das β-Acetyl-α-pyronderivat *180*, während das reaktionsträgere *179* mit dem Enaminteil von *118b* zum γ-Pyron *181* reagiert. Genauso verhält sich in situ erzeugtes Methylketen *182*. Es dimerisiert zu *179* bevor weitere Reaktion eintritt [19].

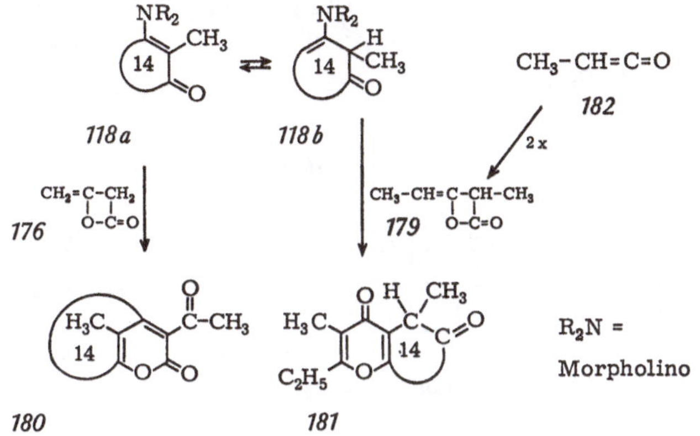

2.5. Acylierung mit Carbonsäurechloriden

Wenngleich bei vielen der beschriebenen Keten-Reaktionen Carbonsäurechloride verwendet wurden, verbleiben noch Acylierungen mit Carbonsäurechloriden, die als solche reagieren, gleichgültig ob ihre Struktur oder die Bedingungen keine Ketenbildung erlauben [73a)].

Aus *Benzoylchlorid* und den entsprechenden Enaminen sind (evtl. in Gegenwart einer Hilfsbase) die Acylenamine *182* [94)] und *183* [95,96)] zugänglich. Hydrolyse von *183* stellt — bei geeigneter Reaktionsführung [67,95,96)] — die beste Methode zur Gewinnung der 2-Benzoyl-cycloanone *184* dar [2)].

R_2N = Morpholino (n=6)
R_2N = Pyrrolidino (n=5)

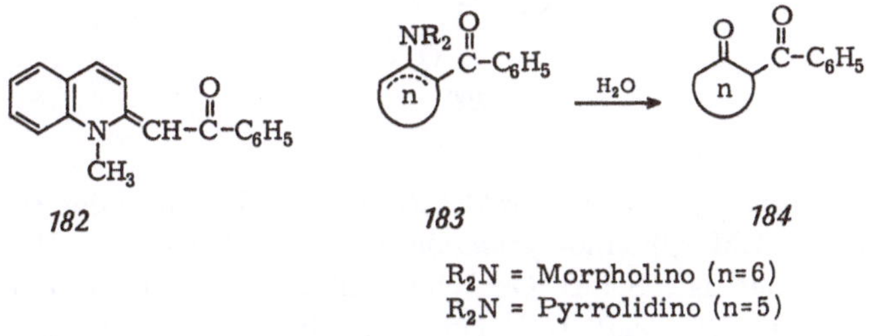

$n = 5{,}6$ NR_2 = Morpholino ($n = 6$), Pyrrolidino ($n = 5$).

Überschüssiges Benzoylchlorid verwandelt *182* und *183* in die Enolester *185* [94] und *186a,b* [67,95]. Die Enolester-Struktur von *186a* sowie des Stereoisomeren *186b* [97] folgt eindeutig aus der Hydrolyse zu den entsprechenden Ketonenolestern [67,95,97], sowie aus dem Auftreten verschiedener Produkte bei succesiver Acylierung des Enamins mit Benzoylchlorid und p-Nitrobenzoylchlorid in vertauschter Reihenfolge [98].

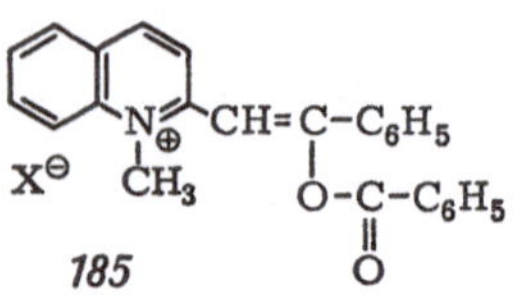

185 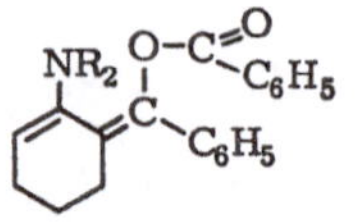   *186a* 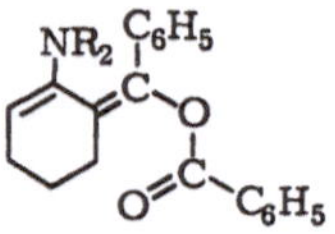   *186b*

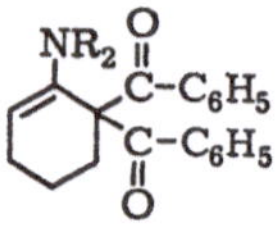

187   *188*   *189*

R_2N = Morpholino

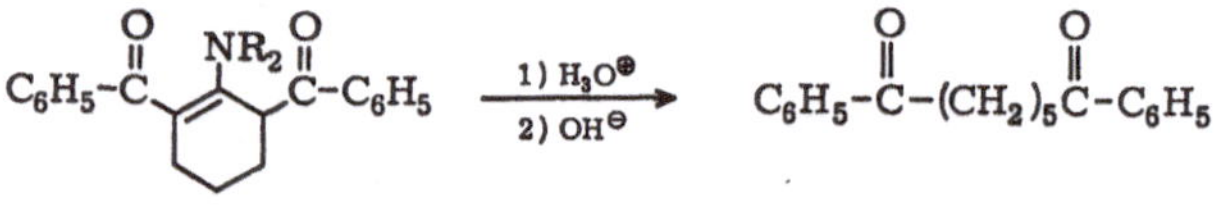

Damit scheiden die Strukturen *187* [99] und *188* aus. Das Isomere *188* ist aus den Ansätzen als Nebenprodukt zu isolieren (Bildung wahrscheinlich über *186a* oder *186b* (vgl. S. 260). Seine Struktur folgt aus der Überführung in 1.5-Dibenzoylpentan (*189*) [95].

Acylierung mit *Phthalolyl-dichlorid* führt sofort zu Enolestern, die im Falle von *190* [95,97] isoliert wurden.

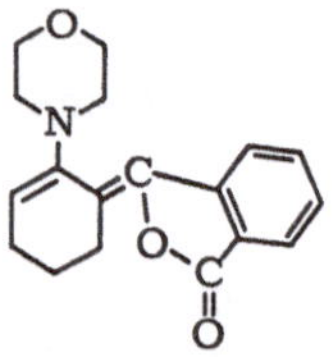

190

Auch *Cyclopropylcarbonsäurechlorid*, das kein Keten bildet [100], acyliert glatt z. B. 1-Morpholino-cyclododecen-1, sodaß nach der Hydrolyse *191* zu gewinnen ist. Selbst Oxalylbromid eignet sich zur doppelten Acylierung, wie *192* zeigt [101]. Durch Hydrolyse entsteht das völlig enolisierte Tetraketon *193*.

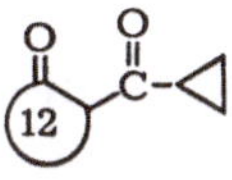

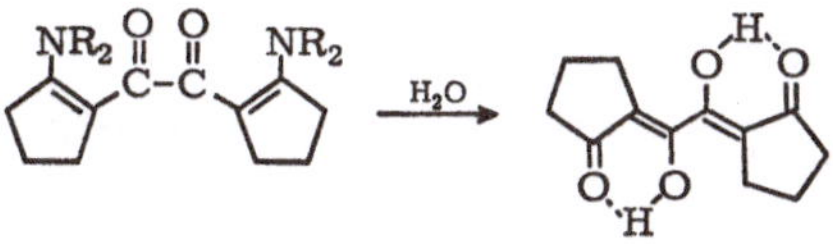

191   *192*   *193*

Auch *Acetylchlorid* kann als solches angreifen. Benutzt man bei der Acetylierung von 1-Morpholino-cyclododecen-1 überschüssiges Enamin als Hilfsbase und nicht Triäthylamin, so unterbleibt die für die Keten-reaktion typische Ringerweiterung und es treten — nach hydrolytischer Aufarbeitung — die „normalen" Produkte *194* und *195* auf [19].

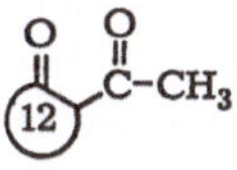

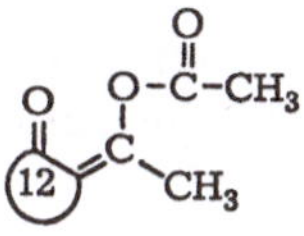

194   *195*

C-Acylierung eines Acylenamins mit Carbonsäurechloriden wurde bisher nur bei der trans-fixierten Verbindung *196* beobachtet, die mit Keten nicht reagiert, während verschiedene Säurechloride *197* erzeugen [102], wobei überschüssiges *196* den entstehenden Chlorwasserstoff unter O-Protonierung zu *198* abfängt [103].

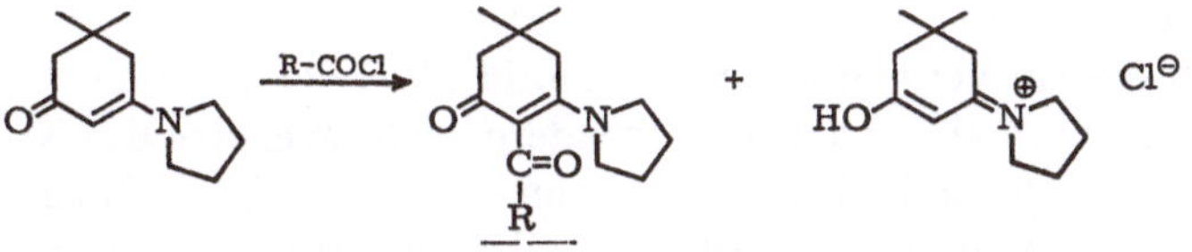

196   *197*   *198*

Da aus der Reaktion von 1-Morpholino(!)-isobuten *46* ($NR_2 =$ Mor-pholino) sowohl mit Acetylchlorid, als auch mit Benzoylchlorid ohne Hilfsbase ein 2.2-Dimethyl-ketoaldehyd (*199*) entsteht, muß auch hier auf direkte Acylierung ohne Ketenzwischenstufe geschlossen werden. Das Addukt *200* ist für $R' = C_6H_5$ isolierbar [104].

$$H_3C \diagdown C \diagup CH_3 \quad \xrightarrow{\text{1 R'-COCl}} \quad CH_3-\overset{\overset{\displaystyle CH_3}{|}}{C}-\overset{\overset{\displaystyle O}{\|}}{C}-R' \quad \xrightarrow{H_2O} \quad O=CH-\overset{\overset{\displaystyle CH_3}{|}}{C}-\overset{\overset{\displaystyle O}{\|}}{C}-R'$$

46	*200*	*199*

$$R' = CH_3, \; C_6H_5$$

2.6. Acylierung mit α,β-ungesättigten Carbonsäurechloriden

Wie die folgenden Beispiele zeigen, kann diese Reaktion ungewöhnlich komplex verlaufen, da neben der normalen Acylierungsreaktion auch Michael-Addition des Enamins an die Doppelbindung erfolgen kann. Diese Alkylierungsreaktion ist z. B. vom Vinylmethylketon oder Acrylnitril wohlbekannt [105]. Sowohl die Natur des Enamins und des Säurechlorids als auch die Reaktionsbedingungen beeinflussen das Reaktionsbild entscheidend [73a].

2.6.1. Acylierung von Aldehyd-Enaminen [106]

Setzt man das Enamin *46a* mit Crotonsäurechlorid (*201b*) oder substituierten Zimtsäurechloriden (*201a*) um, so isoliert man nach Hydrolyse die ungesättigten Ketoaldehyde *203*, also die normalen Acylierungsprodukte. Mit Acrylsäurechlorid (*201c*) hingegen wird statt *203* (R¹ = H) der Ketodialdehyd *209* als Hydrolyseprodukt eines 2:1-Adduktes gefaßt. Mit Crotonsäurechlorid läßt sich diese Addition nur erzwingen, wenn das stärker nucleophile Pyrrolidin-Enamin *46b* eingesetzt wird. Hier machen die gefaßten Zwischenstufen *207* und *208* das Reaktionsschema I wahrscheinlich.

Nach diesem vermag das Säurechlorid *201b* das Enamin *46b* nicht nur zu *202* „normal" zu acylieren, sondern auch reversibel das N-Acylierungsprodukt *204* zu bilden. Dieses kann — wahrscheinlich in einer elektrocyclischen Bindungsverschiebung — sich zu *205* umlagern, das sofort neues Enamin *46b* zu *206* aufnimmt (Teil B in *206*). Außerdem ist damit zu rechnen, daß auch *202* — als α,β-ungesättigtes Keton — noch ein Molekül *46b* zu *206* addiert (Teil A in *206*). Das hypothetische *206* stabilisiert sich bei Raumtemperatur zu *208*, dessen Struktur zusammen mit seinem Hydrolyseprodukt *210* gesichert ist. Bei höherer Temperatur vermag *208* bzw. *206* aus dem Medium ein Proton aufzunehmen und damit das isolierbare Bisimmoniumsalz *207* zu bilden, dessen Hydrolyse schließlich zum Keto-dialdehyd *209* führt.

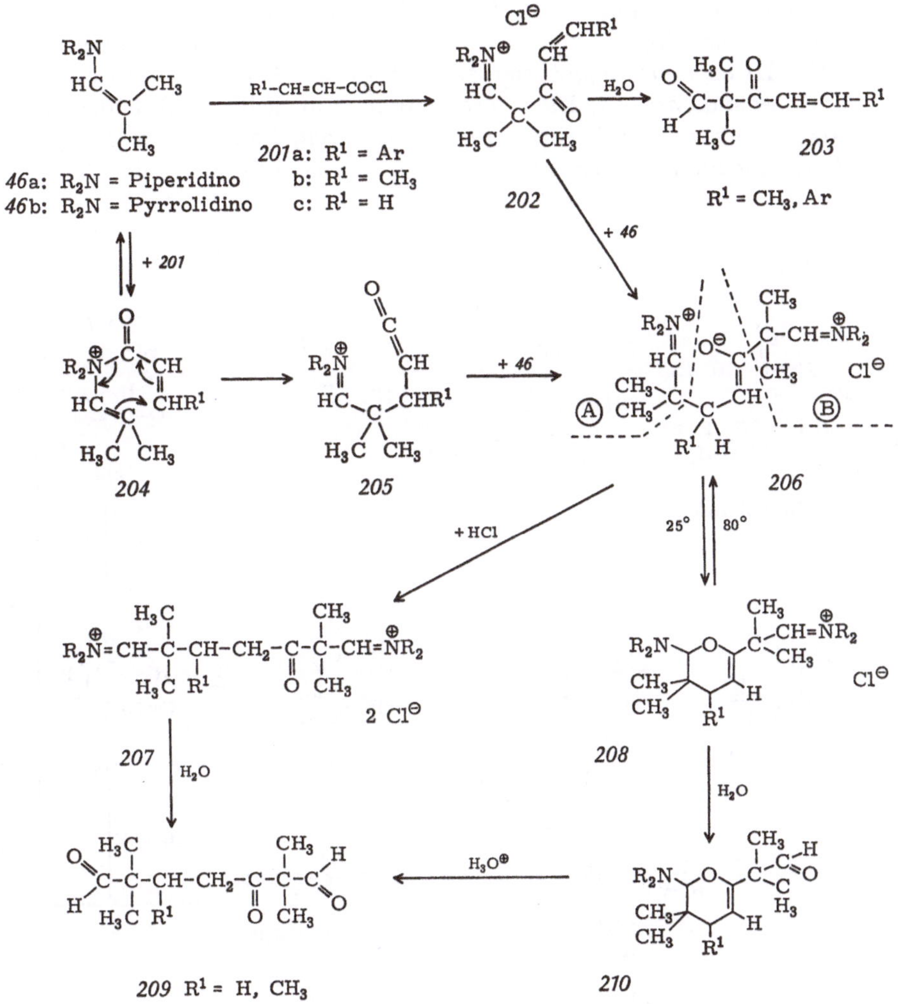

Reaktionsschema I

2.6.2. Acylierung von Ketonenaminen

Keton-Enamine, deren Doppelbindung in zwei isomeren Stellungen auftreten können, bieten eine neue Möglichkeit: Acylierung und Alkylierung durch das α,β-ungesättigte Carbonsäurechlorid können im gleichen Molekül stattfinden, wobei ein *neuer Sechsring* entsteht (Schema II). So reagieren die Säurechloride *211 a,b,c* mit dem Enamin *212 a* in Benzol, vermutlich über *213*, *214* und *215* zu *216*. Dieses ist zum Acylenamin *217* ($R^2 = H$) oder *218* deprotonierbar, oder zu den definiert substituierten Cyclohexandionen-1.3 *219* hydrolisierbar. Auf jeden Fall findet die Alkylierung vor der Acylierung statt, da Äthanol aus einer salzartigen Zwischenstufe (*214*) den δ-Ketoester *220* erzeugt [107].

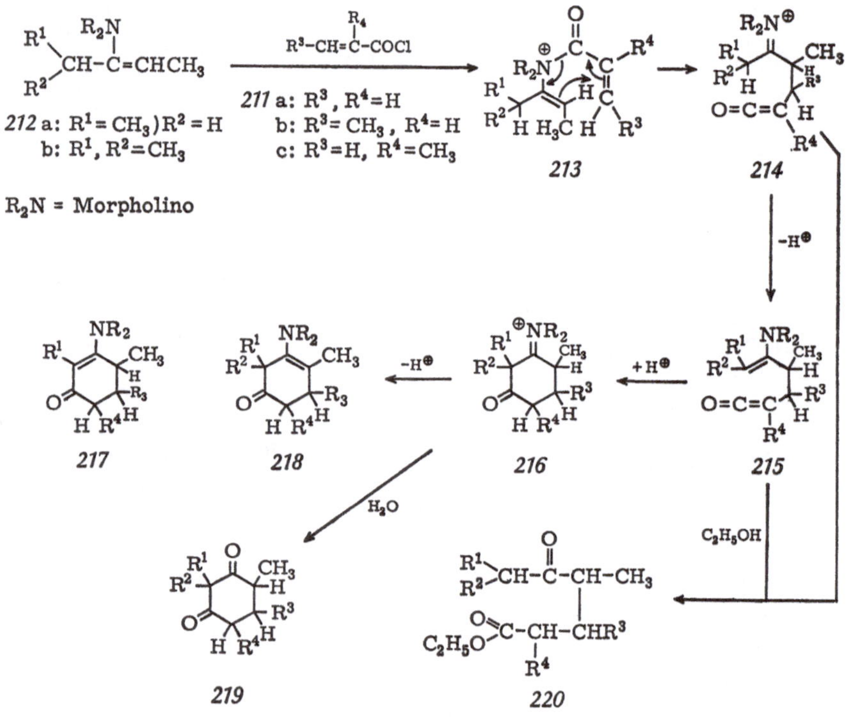

$$R^3\text{-CH=C-COCl}$$

212 a: $R^1 = CH_3$) $R^2 = H$
b: $R^1, R^2 = CH_3$

211 a: R^3, $R^4 = H$
b: $R^3 = CH_3$, $R^4 = H$
c: $R^3 = H$, $R^4 = CH_3$

R_2N = Morpholino

213 *214* *215* *216* *217* *218* *219* *220*

C_2H_5OH

H_2O

$-H^\oplus$ $+H^\oplus$

Schema II

Die Enamine cyclischer Ketone schließen sich in ihrem Verhalten an, wie die Reaktion von *93* mit Acrylsäurechlorid zeigt. Man isoliert das Enaminketon *221*, dessen Hydrolyseempfindlichkeit stark von der Ringgröße abhängt (Bredt'sche Regel!), oder das Diketon *222*. Im Falle von *93* $n = 6$ wurde bei der Hydrolyse auch das 2:1 Adduktt *223* gefunden.

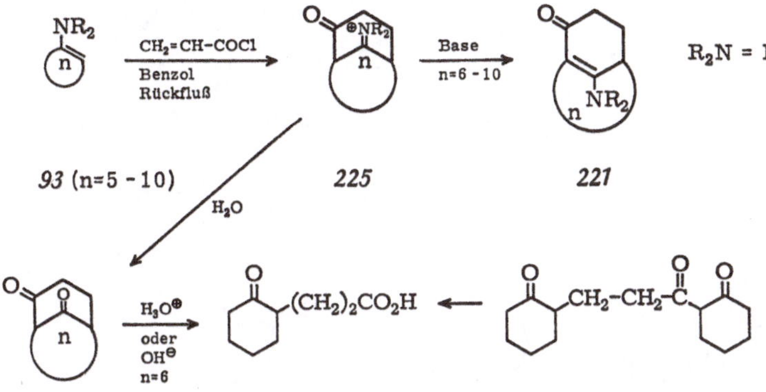

93 (n=5 - 10)　　　　　　　　*225*　　　　　　*221*

222 (n=6 - 8)　　　*224*　　　　　　*223*

224 tritt nur als Sekundärprodukt durch Spaltung von *223* oder *222* auf. Damit ist nochmals die Acylierung als Primärschritt sichergestellt. Selbst mit Zimtsäurechlorid läßt sich in siedendem Benzol eine Reaktion zu phenyl-substituierten Produkten der Formel *222* und *223* erzwingen [109]. Dagegen setzt sich das Enamin *226* in Äther zu den „normalen" Acylierungsprodukten *227* oder *228* um [19].

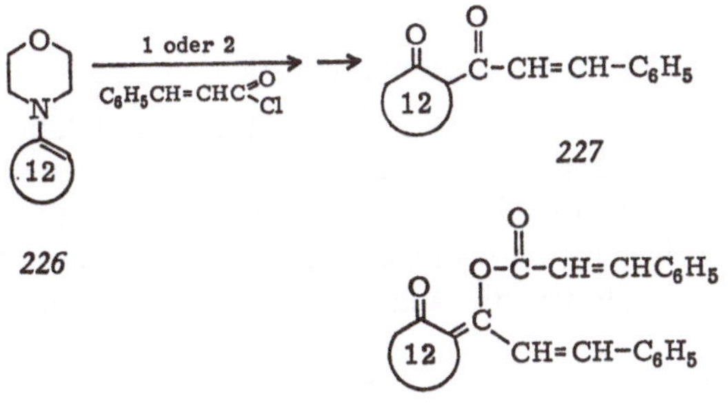

226　　　*227*

228

2.7. Acylierung mit vinylogen Carbonsäurechloriden

Durch formalen Austausch von Chlor und β-Substituent entstehen aus α,β-ungesättigten Carbonsäurechloriden die Chlorvinylketone *229*, die sich ähnlich wie Säurechloride verhalten. Ihre Reaktionen seien daher hier besprochen, obwohl es sich um Alkylierungen handelt. Die Enamine *34* reagieren relativ glatt zu den erwarteten Produkten *230*, welche infolge ihres Merocyanin-Charakters gelbe bis rote Farbe zeigen [110a].

34 229 230

232 233 231

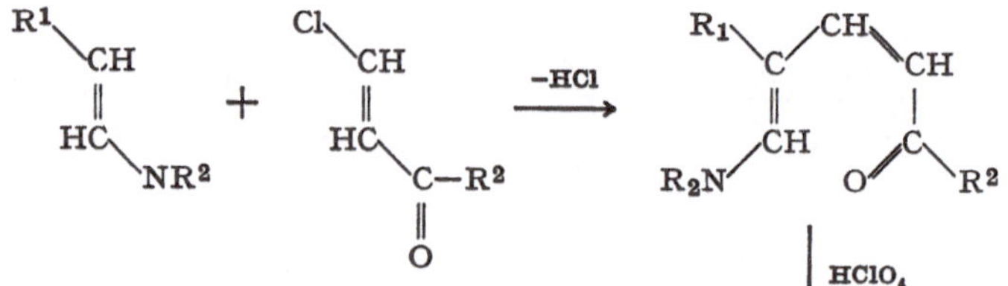

$R^1 = CH_3, C_2H_5, C_6H_5$
$R^2 =$ Aryl (Alkyl)
$R_2N =$ Pyrrolidino, Piperidino, Morpholino.

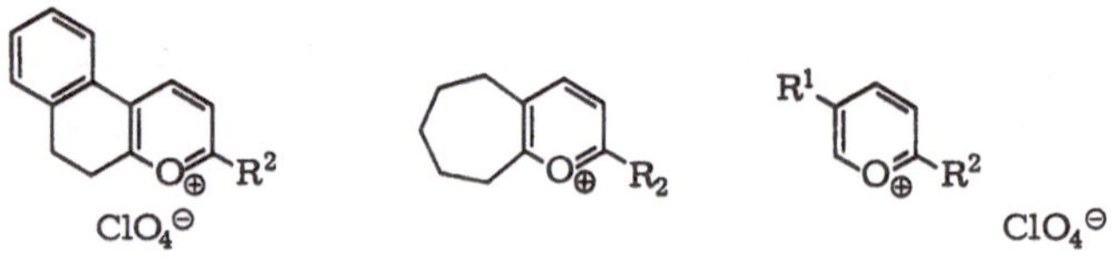

Der Wert dieser Reaktion besteht darin, daß *230* mit Perchlorsäure — häufig quantitativ — in die sonst schwer zugänglichen 2.5-disubstituierten Pyryliumsalze *231* übergeht [110b]. Mit geeigneten Ausgangsenaminen sind auf diese Weise z.B. auch *232* und *233* zugänglich.

2.8. Acylierung mit Carbonsäureanhydriden

Dieser Methode kommt wenig Bedeutung zu. Zwar läßt sich 1-Pyrrolidinocyclohexan-1 (*234*) mit Acetanhydrid mit 42% Ausbeute in 2-Acetylcyclohexanon (*235*) verwandeln [38], doch wirkt Acetylchlorid effektiver [111]. Wichtiger ist die Bildung von *236* mit Ameisensäure-essigsäureanhydrid [38], doch ist die Vilsmeier-Variante vorzuziehen. Dagegen

235 234 236

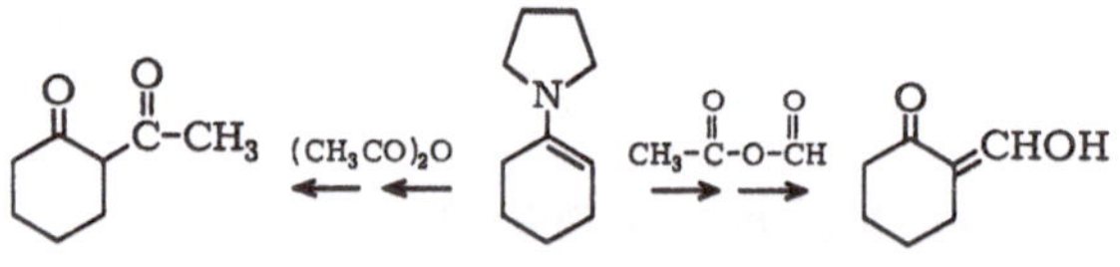

erlaubt das gemischte Anhydrid *237* aus Acetyl-salicylsäure die Darstellung von Chromonen (*238*) unter Umgehung der für die Synthese von γ-Pyronen hier nicht möglichen Diketene [112] (vgl. S. 264).

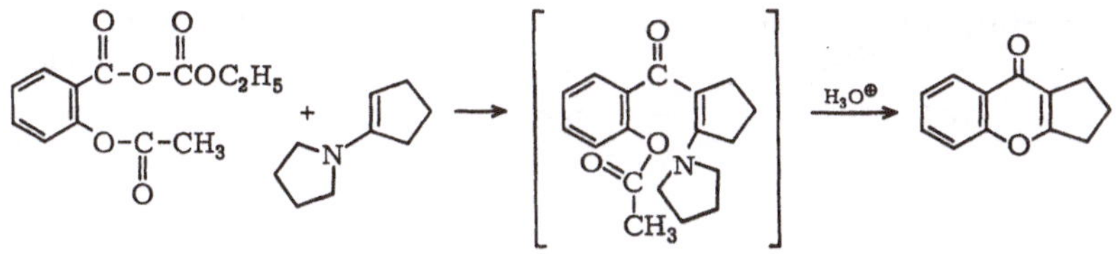

237   238

2.9. Acylierung mit Vilsmeier-Reagentien

An Stelle der Claisen-Kondensation von Ketonen mit Ameisenestern oder
der direkten Vilsmeier-Formylierung lohnt sich zur Gewinnung von
α-Hydroxymethylen-ketonen 239 der Umweg über die Enamine 240. Die

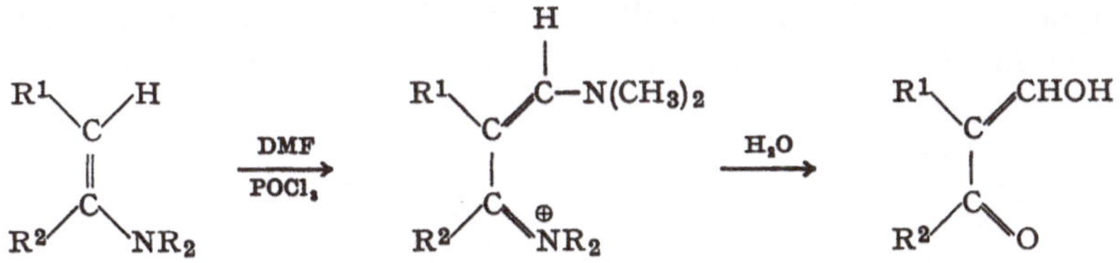

240   241   239

cyaninartige Zwischenstufe 241 [112a)] wird meist nicht isoliert [113,114)]. In
analoger Weise sind aus 46 die Bis-Immonium-Salze 242 in hohen Ausbeuten darstellbar [115)].

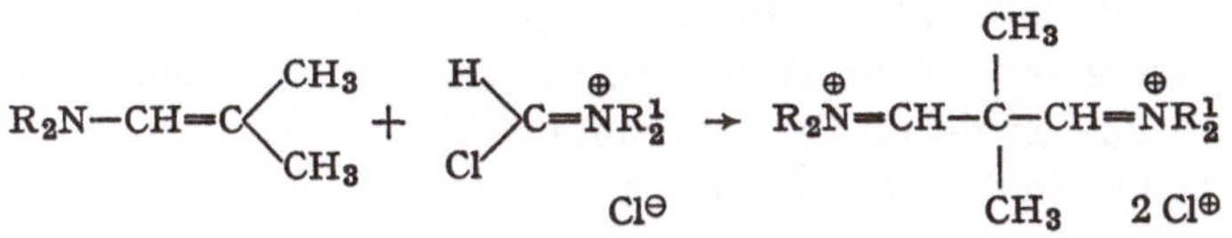

46   242

Auch der Befund, daß die Dialkylamide der höheren Carbonsäuren
243 bei der Überführung in Vilsmeier-Reagentien 244 häufig „Dimerisa-
tion" zu 245 bzw. 246 erleiden, ist als Acylierung des Enamins 247 durch
244 zu verstehen [116)].

273

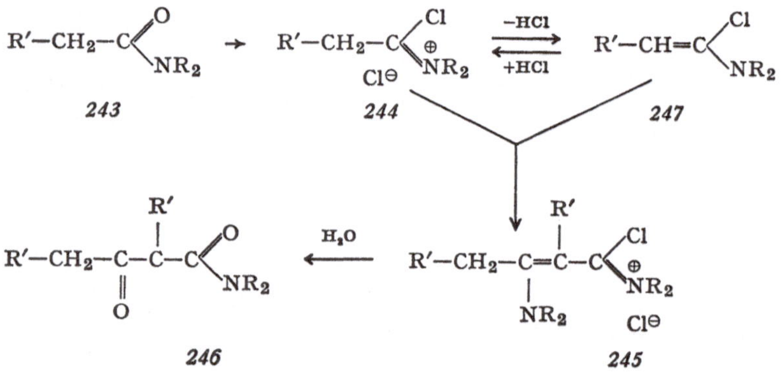

2.10. Acylierung mit Imidosulfonaten

Imidsäurederivate *248* sollten Enamine ebenfalls acylieren, sofern X stark nucleofug ist. Diese Variante bietet neue präparative Möglichkeiten, wenn die entsprechenden Lactim-Derivate *249* eingesetzt werden, wegen ihrer Solvolyseempfindlichkeit zweckmäßig durch Beckmann-Umlagerung in situ erzeugt. Dies geschieht am besten durch Überführung der

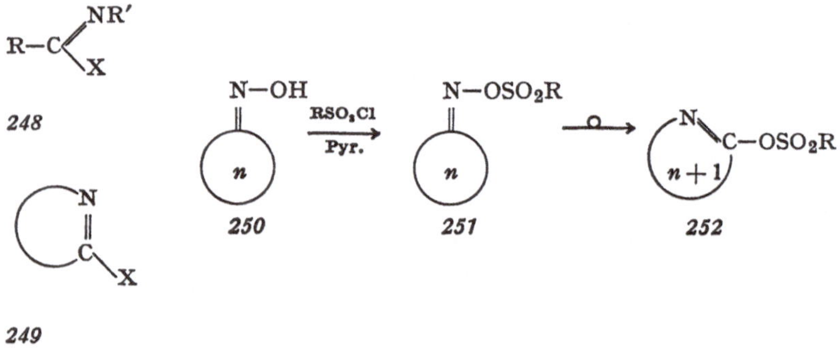

Oxime *250* in ihre Sulfonsäureester *251*, die schon bei Raumtemperatur zum Lactimester *252* umlagern [117]. Der aus Cyclohexanon dargestellte Caprolactimester *253* reagiert glatt mit den Enaminen *93*, wobei zunächst das cyaninartige Substitutionsprodukt *254* anfällt [118]. Die für ein Trimethincyanin sehr langwellige Absorption, von $n = 5$ bis $n = 10$ zunehmend, wird wahrscheinlich, wie in anderen Fällen [119,120], durch Verdrillung des Chromophors hervorgerufen. Der relativ niedrige, von n fast unabhängige Extinktionskoeffizient deutet in die gleiche Richtung. Deprotonierung von *254* mit starken Basen erhöht den Energieinhalt so stark, daß — soweit untersucht [121] — neben *255* das sterisch weniger gespannte, nichtkonjugierte Isomere *256* auftritt (UV-Bande des Enamins, Vinyl-H im IR- und NMR-Spektrum).

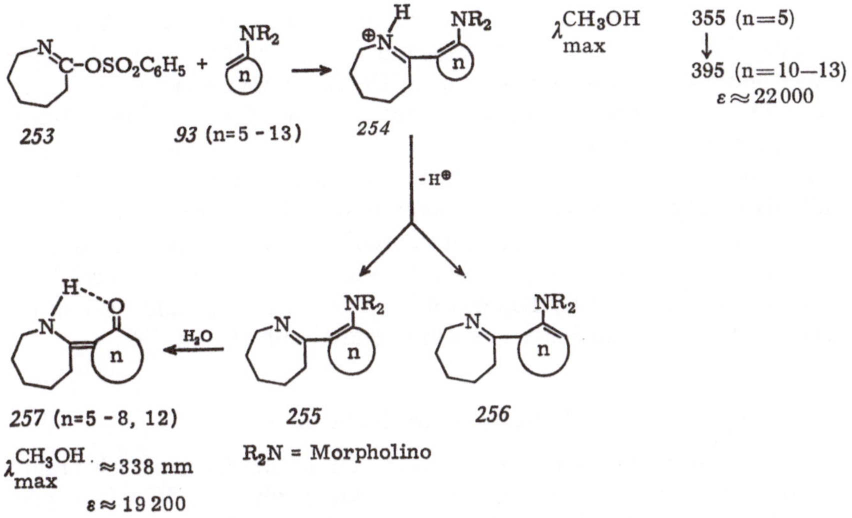

$$253 \qquad 93\ (n=5-13) \qquad 254$$

$$\lambda_{max}^{CH_3OH} \qquad 355\ (n=5)$$
$$395\ (n=10-13)$$
$$\varepsilon \approx 22\,000$$

$$-H^{\oplus}$$

$$257\ (n=5-8,\,12) \qquad\qquad 255 \qquad\qquad 256$$

$$\lambda_{max}^{CH_3OH} \approx 338\ nm \qquad R_2N = Morpholino$$
$$\varepsilon \approx 19\,200$$

Milde Hydrolyse überführt *255* in *257*, das auf Grund von spektro-skopischen Daten als H-verbrücktes Enaminoketon zu formulieren ist.

Eine ganz analoge Reaktionsfolge läßt sich durch Variation der Ring-größe des Lactimesters *252* verwirklichen [121)].

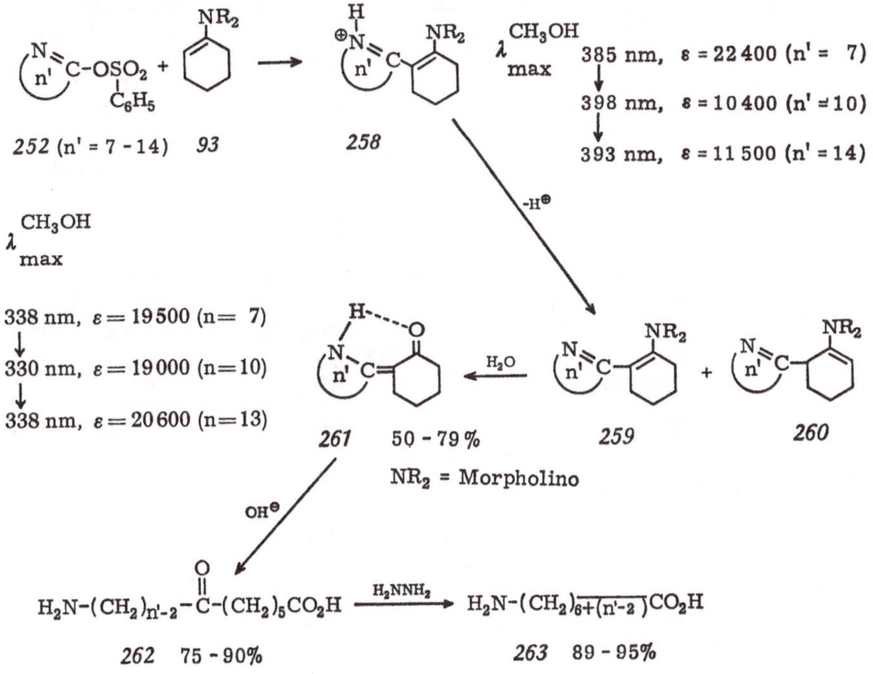

$$252\ (n' = 7-14) \qquad 93 \qquad\qquad 258$$

$$\lambda_{max}^{CH_3OH}$$
$$385\ nm, \quad \varepsilon = 22\,400\ (n' = 7)$$
$$398\ nm, \quad \varepsilon = 10\,400\ (n' = 10)$$
$$393\ nm, \quad \varepsilon = 11\,500\ (n' = 14)$$

$$-H^{\oplus}$$

$$\lambda_{max}^{CH_3OH}$$

$$338\ nm, \quad \varepsilon = 19\,500\ (n = 7)$$
$$330\ nm, \quad \varepsilon = 19\,000\ (n = 10)$$
$$338\ nm, \quad \varepsilon = 20\,600\ (n = 13)$$

$$261 \quad 50-79\,\% \qquad\qquad 259 \qquad\qquad 260$$

$$NR_2 = Morpholino$$

$$OH^{\ominus}$$

$$H_2N-(CH_2)_{n'-2}-\overset{O}{\overset{\|}{C}}-(CH_2)_5 CO_2 H \xrightarrow{H_2NNH_2} H_2N-(CH_2)_{\overline{6+(n'-2)}} CO_2 H$$

$$262 \quad 75-90\,\% \qquad\qquad\qquad 263 \quad 89-95\,\%$$

Kondensation mit *93* ($n = 6$) führt zu *258* bei dem — umgekehrt wie bei *254* (S. 275) — die Lage von λ_{max} nur wenig, die Intensität der Bande aber stark von n abhängt. Das beim Deprotonieren erhaltene Gemisch von *259* und *260* besteht für $n' = 7$ fast ganz aus *260* [121]. Die hydrolytisch erzeugten Enamino-ketone *261* ändern ihre spektralen Eigenschaften — genau wie *257* (S. 275) — nur unbedeutend mit der Ringgröße. Die alkalische Hydrolyse von *261* erfolgt praktisch quantitativ am Cyclohexanonring zu den ω-Amino-ketosäuren *262*, die sich glatt zu den ω-Aminosäuren *263* reduzieren lassen. Damit liegt ein bequemes Verfahren vor, um letztlich aus Cyclanonen mit $n = 6{-}13$ ($\rightarrow 252$) und Cyclohexanon ($\rightarrow 93$, $n = 6$) langkettige ω-Aminosäuren aufzubauen [122].

2.11. Acylierung mit Hydroxamsäurechloriden [123]

Bei dieser Reaktion, die in Gegenwart von Triäthylamin durchgeführt wird, handelt es sich nicht um eine direkte Acylierung. Vielmehr geht das Hydroxamsäurechlorid *264* zunächst in ein Nitriloxid *265* über, das sich als 1.3-Dipol an das Enamin *240* addiert. Das in vielen Fällen

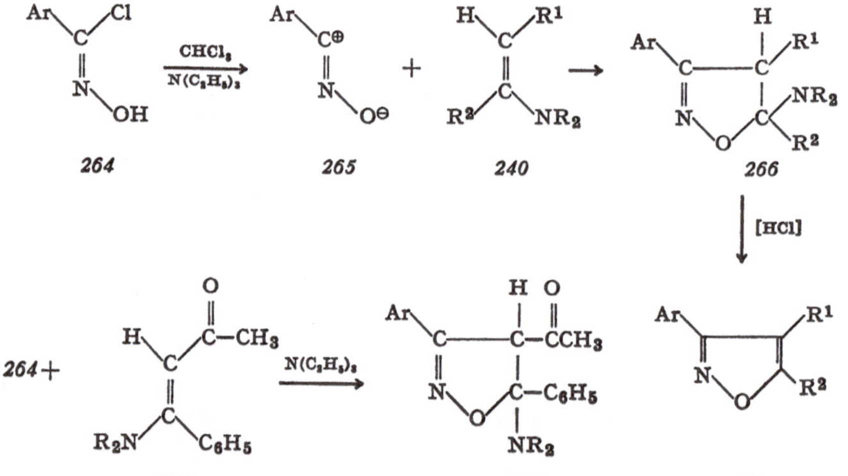

R^1 = H, Alkyl

R^2 = H, Alkyl
        Aryl

NR$_2$ = Piperidino
         Morpholino
         Methylanilino

isolierbare Addukt *266* ist glatt in das Isoxazol *267* überführbar. In Substanz eingesetzte Nitriloxide verhalten sich ebenso. Diese Cycloaddition macht verständlich, daß auch Acylenamine, wie z. B. *268*, mit Hydroxamsäurechloriden zu Addukten (*269*) reagieren, die sich ebenfalls zu Isoxazolen stabilisieren.

2.12. Acylierung mit Hydrazidchloriden [124)

Auch diese Reaktion ist als 1.3-dipolare Addition der aus z. B. *270* erzeugten Nitrilimine *271* an Enamine (z. B. *93*) zu verstehen. Die Addukte *272* gehen für $n = 6$ unmittelbar in *273* über, während sie mit $n = 5$ zu fassen sind. Setzt man heterocyclische Enamine, wie z. B. *274* ein, so entstehen Pyrazole mit einer ω-Aminoalkylgruppe in 4-Stellung (*275*). Weitere 1.3-dipolare Additionen an Enamine vgl. [125).

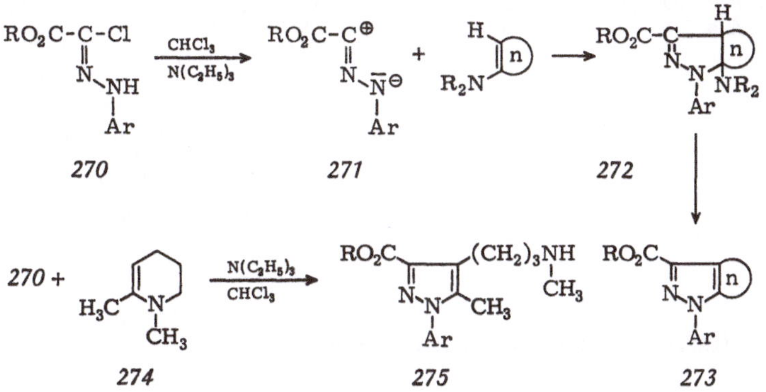

2.13. Acylierung mit vinylogen Amidinium-Salzen

Das vinyloge Amidiniumion *276* vermag das Enamin *2* zu *277* und *278* zu acylieren. *277* und *278* sind mono- bzw. di-cis-fixierte Cyanine [126).

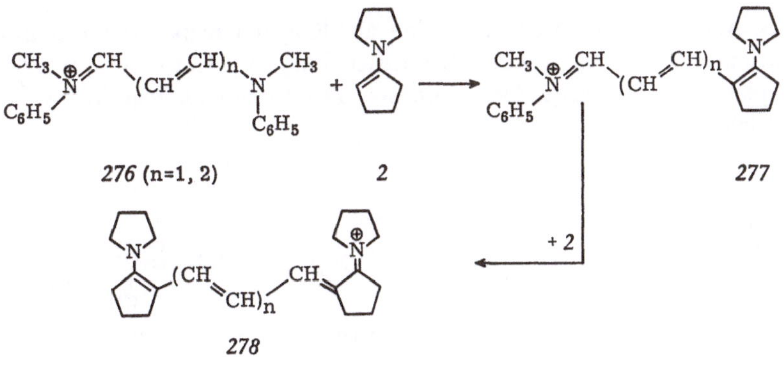

2.14. Gekoppelte Enaminbildung und Acylierung

Mit dieser originellen Methode ist es möglich, sogar die Carbonester- und Nitrilfunktion zur Acylierung des Enamins einzusetzen.

Aus *279* bzw. *280* und Cycloanonen entstehen die Chinolizinderivate *281* und *282* [127,127a].

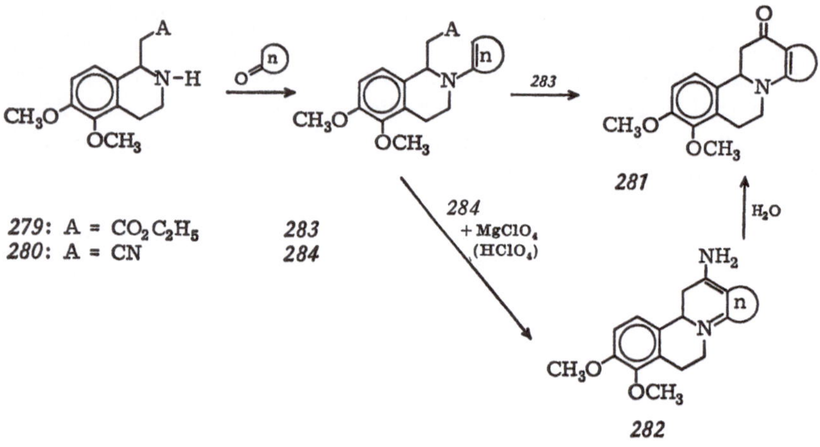

279: A = $CO_2C_2H_5$
280: A = CN

283
284

3. Acylierung mit Kohlensäurederivaten

Die Reaktionen von Kohlensäurederivaten mit Enaminen weisen einige Besonderheiten auf. Aber auch hier beobachtet man „normale" Acylierung, bei der Carbonsäurederivate entstehen, sowie Cycloadditionen.

3.1. Acylierung mit Kohlensäureesterchloriden

Diese Acylierung gehört mit zu den am längsten bekannten Enaminreaktionen [38]. Sie bietet einen bequemen Weg zu β-Ketoestern, wie das Beispiel *130→284* zeigt. Der Ketoester *285* ist auf analoge Weise zugänglich [38].

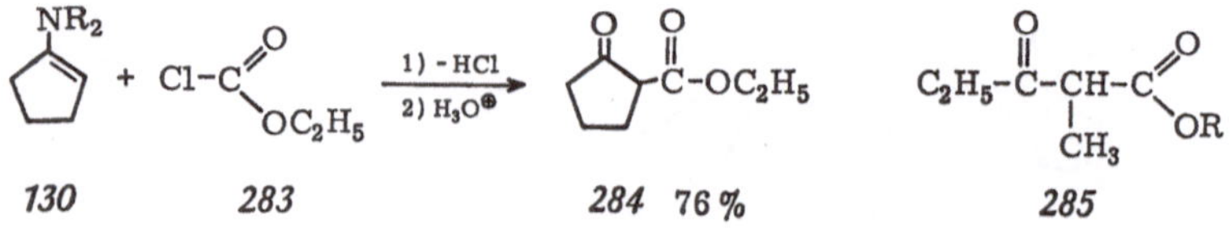

130   283   284 76 %   285

Triäthylamin ist hier als Hilfsbase unbrauchbar [38)], da das mit dem Esterchlorid 283 rasch entstehende Quartärsalz zu reaktionsträge ist [128)]. Man muß also entweder ein zweites Mol Enamin [38)] oder das sperrige Äthyl-diisopropylamin als Hilfsbase benutzen [129)].

3.2. Acylierung mit Orthokohlensäurederivaten

Erstaunlicherweise wirkt selbst *Tetrachlorkohlenstoff* bei längerer Reaktionszeit als Acylierungsmittel [130)]. Nach hydrolytischer Aufarbeitung fallen die Dichlor-methylketone 286 an während aus Dimethylamino-isobuten der β-Trichlormethylaldehyd 286a (70%) entsteht [130a)].

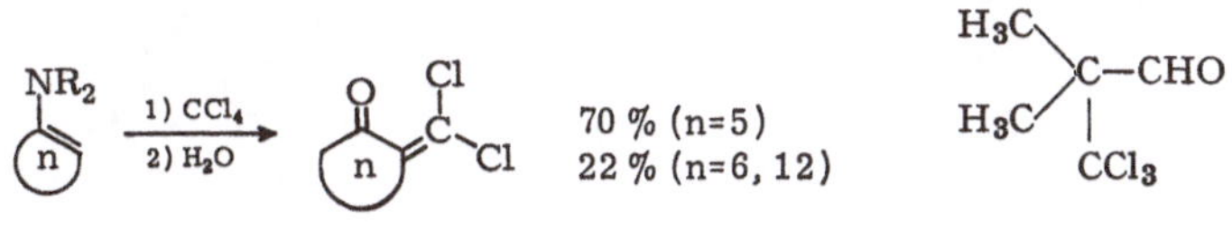

93 (n=5, 6, 12)   286   286a

R$_2$N = Morpholino

Mit Hilfe des Immoniumsalzes 287, dargestellt aus dem Ortho-kohlensäure-dichlorid 288 und Triäthylamin, sind ebenfalls β-Ketosäure-derivate zugänglich, wie die Reaktion 287 + 93 (n = 6) → 290 zeigt [131)].

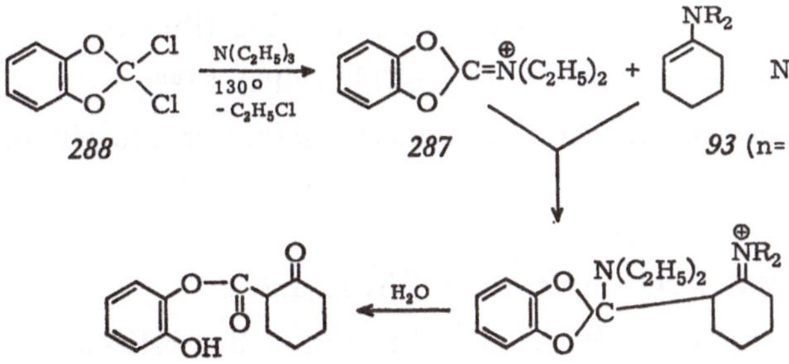

288   287   93 (n=6)

290   289

3.3. Acylierung mit Cyansäurederivaten

Für die Reaktion des Chlorids der Cyansäure mit einem Enamin [73a)] kann wieder die Fischer-Base 4 als Musterfall dienen [132)]. Ketonenamine 93 führen zu 2-Cyan-cyclanonen 292 [133)].

Nachdem Arylcyanate bequem zugänglich geworden sind, dürften sich diese als Acylierungsmittel empfehlen [134,135]. Im Brom- und Jodcyan ist die Polarität vertauscht: Das Halogen tritt in die 2-Stellung des Enamins ein [136]. Cyanurchlorid acyliert erwartungsgemäß zu 293 [137].

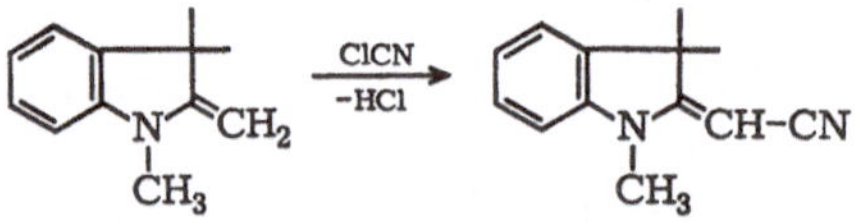

n=5 - 9 R_2N = Morpholino
n=5, 6 R_2N = Morpholino, Piperidino, Pyrrolidino

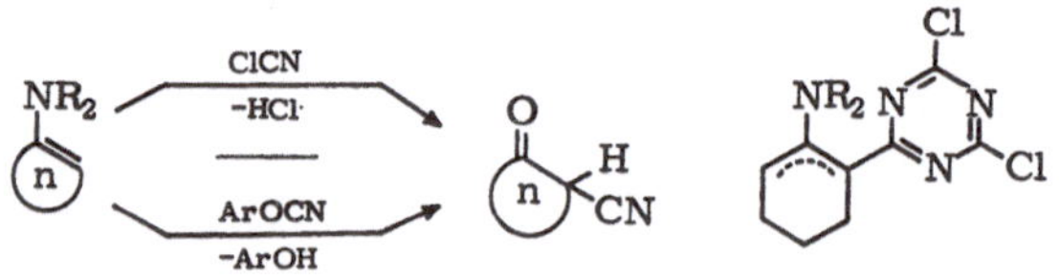

3.4. Acylierung mit Isocyanaten und Isothiocyanaten [45a]

Diese Reaktion ähnelt in vieler Hinsicht derjenigen mit Ketenen, doch wird auch auf prinzipielle Unterschiede hinzuweisen sein. Die Elektrophilie der Isocyanate steigt in der Reihenfolge der Substituenten

Alkyl < Aryl < Aroyl < Arylsulfonyl [138] < Chlorsulfonyl [139];

für die grundsätzlich schwächer elektrophilen Isothiocyanate gilt das Gleiche [140].

3.4.1. Acylierung von β-substituierten Aldehydenaminen

Aldehydenamine wie 46, in denen die β-Stellung voll substituiert ist, verbinden sich mit Phenylisocyanat fast quantitativ zum Cycloaddukt 294, das sich zum β-Formyl-anilid 295 hydrolysieren läßt [141,143]. Daß der Ringschluß des Zwitterions 296 reversibel ist, wird durch die Reaktion mit überschüssigem Phenylisocyanat bei höherer Temperatur zum 2:1 Addukt 297 wahrscheinlich gemacht [143]. Die Zwischenstufe 296 ist außerdem bei der Dreikomponentenreaktion mit Isonitril zu 298 zu fordern [144]. Hier lassen sich sogar Isothiocyanate einsetzen [144], von denen bisher noch keine 1:1 Cycloaddukte bekannt sind.

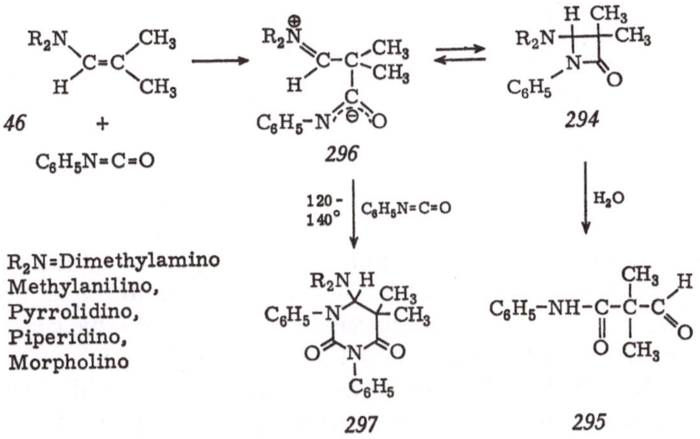

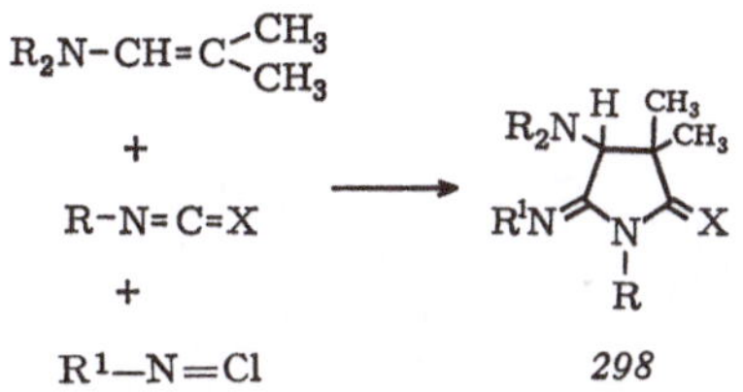

$$R_2N \diagup C=C \diagdown \begin{smallmatrix} CH_3 \\ CH_3 \\ CH_3 \end{smallmatrix} \longrightarrow R_2\overset{\oplus}{N} \diagup C-C \diagdown \begin{smallmatrix} CH_3 \\ CH_3 \end{smallmatrix} \rightleftharpoons$$

46   +

$C_6H_5N=C=O$

296

294

R_2N = Dimethylamino
Methylanilino,
Pyrrolidino,
Piperidino,
Morpholino

$\begin{array}{c} 120- \\ 140° \end{array} \Big| C_6H_5N=C=O$ $\Big| H_2O$

297

295

$$R_2N-CH=C \diagdown \begin{smallmatrix} CH_3 \\ CH_3 \end{smallmatrix}$$

+

$R-N=C=X$

+

$R^1-N=Cl$

298

Pyrrolidino,
R_2N = Piperidino,
Morpholino

R, R^1 = Alkyl, Cycloalkyl
Aryl

X = O, S

3.4.2. Acylierung von Aldehyd- und Ketonenaminen mit β-ständigem H-Atom

Sobald das Enamin über β-ständigen Wasserstoff verfügt (z. B. *240*), sind trotz starker Variation der Enamin- und Iso-(isothio)cyanat-Komponente 299 nur lineare Addukte *300* zu fassen [145,146]. Das gilt in gleicher Weise für die Addukte *301* und *302* aus heterocyclischen Methylbasen [147,148].

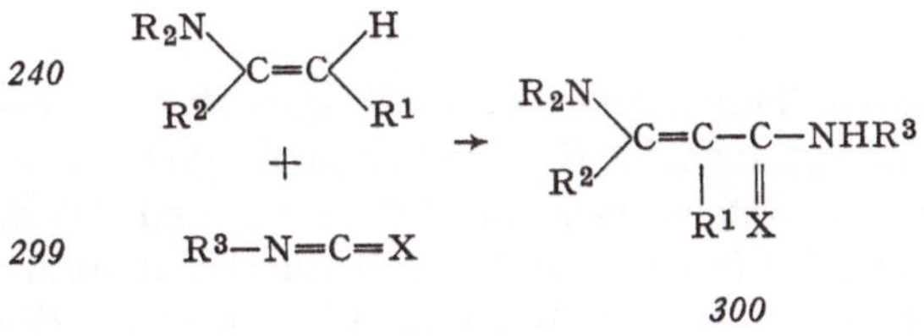

240

$$R_2N \diagup C=C \diagdown \begin{smallmatrix} H \\ R^1 \end{smallmatrix}$$

+

299 $R^3-N=C=X$

$\rightarrow$

$$R_2N \diagup C=C-C-NHR^3 \atop R^2 \diagdown \qquad R^1 \; X$$

300

R_2N = Pyrrolidino,   R^1 = H, Alkyl, C_6H_5   R^3 = Alkyl, C_6H_5,
Morpholino,   R^2 = H, Alkyl   ClSO_2
Acetyl-methylamino,   R^1 + R^2 = (CH_2)_n (n = 3,4,5)   X = O,S

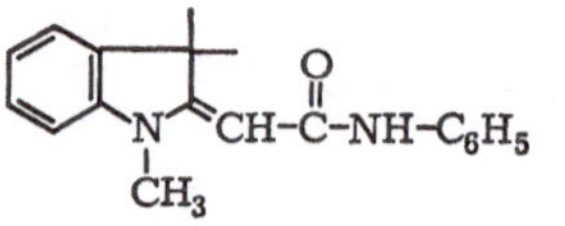

301 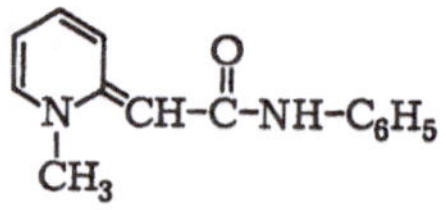 302

Der Versuch, bei der Reaktion von 1-Morpholino-cyclododecen-1 mit *Phenylisocyanat* in Heptan bei 0 °C ein Cycloaddukt vom Typ *294* (S. 281) als Zwischenstufe nachzuweisen, verlief negativ, die charakteristische Bande bei 1730—1750 cm^{-1} [141,142)] trat nicht auf.

Die hohe Nucleophilie der C=S-Bindung im Thioamid macht die Thioisocyanat-Addukte *300* zu wertvollen Ausgangsmaterialien für Heterocyclen-Synthesen, z.B. *303→304* und *305→306* [149)].

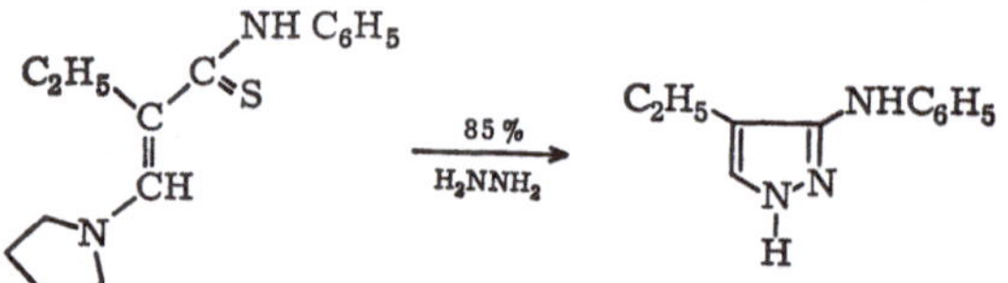

303 304

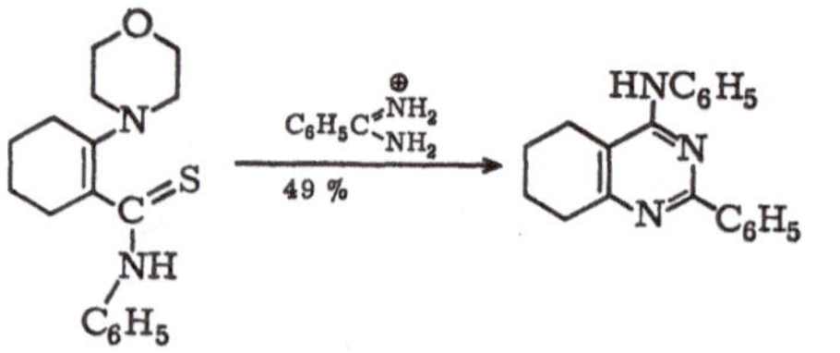

305 306

Eine Sonderstellung nehmen *Acylisothiocyanate* als Acylierungsmittel ein, da die Aminogruppe des Enamins unter Ringschluß verdrängt werden kann. So entsteht zwar aus *307* ($n=5$) und *308* das erwartete Addukt *309*, mit *307* ($n=6$) und *308* wird jedoch unmittelbar *310* erhalten, ein Derivat des Aza-γ-thiopyrons, das sich über *311* glatt in das Pyrimidinderivat *312* verwandeln läßt [149)]. Das Pyrimidin-System kann auch direkt aus *307* mit dem amidinartigen *313* aufgebaut werden, wie Beispiel *314* zeigt [150,151)].

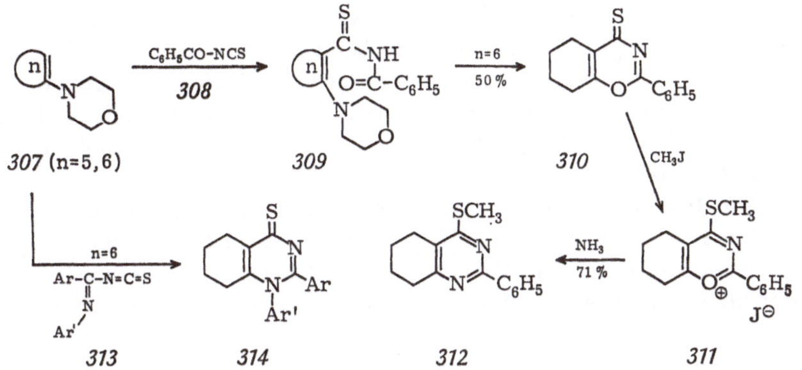

3.4.3. Zweifache Acylierungen isomerisierbarer Enamine

Wie gezeigt, überwiegen bei den Acylierungsprodukten aus Ketonenaminen und Säurechloriden oder Ketenen häufig die Isomeren mit nichtkonjugierter Doppelbindung (s. Kap. 2.3.1.). Ganz entsprechend läßt sich aus dem Adduktgemisch *315 + 316* (in 93% Ausb. aus *307* ($n = 6$) und Phenylisocyanat erhalten) durch Umkristallisieren aus Acetonitril das Isomere *316* mit 60% Ausbeute abtrennen, das sich durch Vinyl-H (NMR) und Fehlen einer langwelligen Bande bei ≈ 300 nm verrät [146a, 152]. Ein zweites Molekül Isocyanat wird daher zu *317* aufgenommen, dessen Struktur durch unabhängige Synthese seines Hydrolyseproduktes *318* gesichert ist [146a,152,138a]. Mit Phenylisothiocyanat entsteht erwartungsgemäß *319*, das sich zu *320* hydrolysieren und schließlich durch Entschwefeln mit H_2O_2 in *318* überführen läßt [152].

Die Addition an die CO- oder NH-Funktion [153] der Amidgruppe unterbleibt in beiden Fällen. Im Isothiocyanat-Addukt an das gleiche

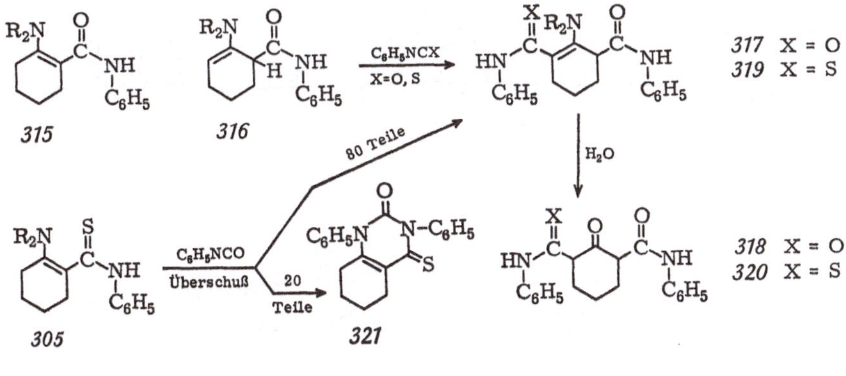

$R_2N = $ Morpholino

Enamin scheint das konjugierte *305* vorzuherrschen, wie bereits seine gelbe Farbe zeigt [138a]. Dementsprechend addiert sich Phenylisocyanat nicht zu *319*. Vielmehr verdrängt es unter verschärften Bedingungen den Isothiocyanat-Rest, so daß *317* entsteht. Daneben greift es noch an der NH-Funktion der Thioamidgruppe an, denn man isoliert zusätzlich das Thiouracil *321* [152,145b].

Die beschriebenen Reaktionen sind auch am 1-Morpholino-cyclopenten-1 beobachtet worden [138a,152,145b]. Aus 1-Pyrrolidinocyclopenten-1 und Aethylisocyanat ist sogar das Bis-addukt *322* erhältlich, das mit überschüssigem Phenylisocyanat *beide* Seitenketten austauscht [152].

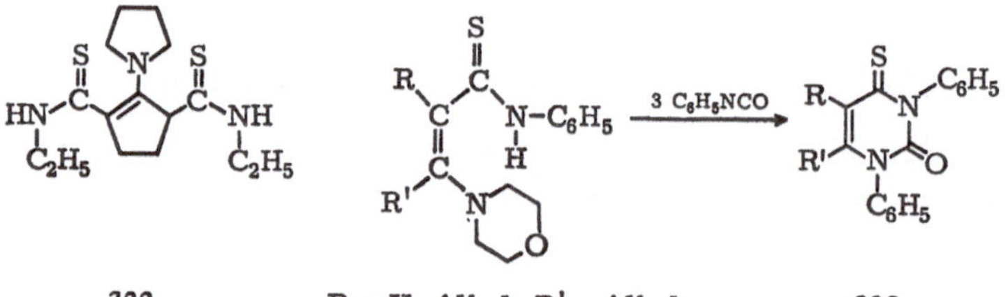

322   R = H, Alkyl R' = Alkyl   *323*

Die Thiouracil-Bildung *323* ist an zahlreichen Beispielen beobachtet worden [145b]. Auf Grund der beschriebenen Reaktionen entspricht es den Erwartungen, daß auch das Acetylierungsprodukt des 1-Pyrrolidinocyclohexens in seiner nichtkonjugierten Form (*115b* mit Phenylisocyanat zu *324* reagiert, das sich zu *325* hydrolysieren läßt [68]. Dagegen war die glatte C-Acylierung der konjugierten Acylenamine *326* durch Arylisothiocyanate zu *327* nicht ohne weiteres vorherzusehen [154].

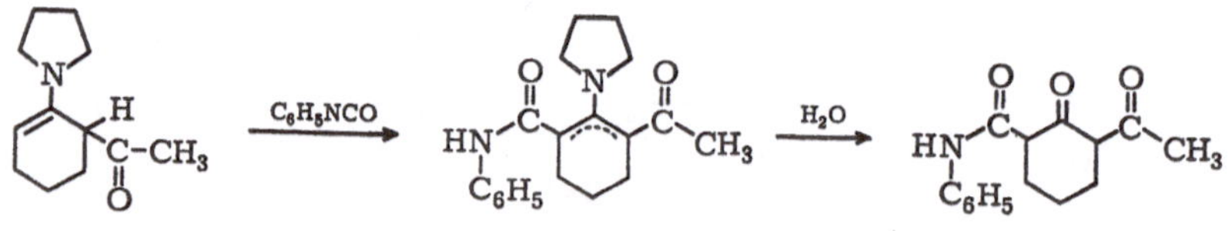

315b   *324*   *325*

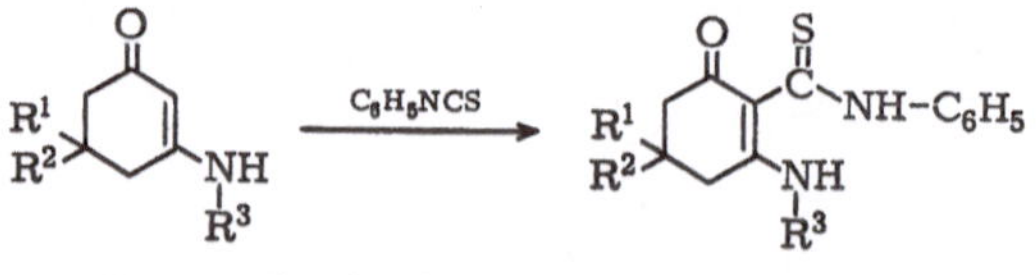

326   R¹, R², R³ = H, CH₃   *327*

3.5. Acylierung mit Schwefelkohlenstoff

Daß die Enamingruppierung der Fischer-Base durch Schwefelkohlenstoff quantitativ zu *328* acyliert wird, ist schon lange bekannt [155]. Überträgt man die Reaktion auf Aldehydenamine *34*, so isoliert man statt der Dithiosäure 3.5-Dialkyl-α-dithiopyrone *330* [156]. Bricht man die Reaktion vorzeitig ab, so findet man die aus den empfindlichen Enaminen *34* durch Kondensation (*34* + *34*) entstandenen Dienamine *329* (vgl. S. 242) als Vorstufe [156].

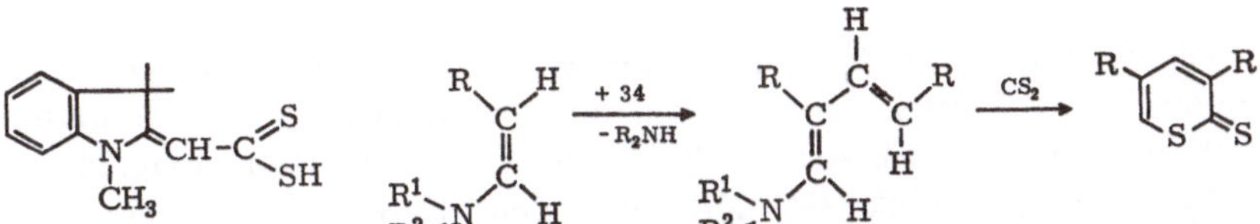

328   *34*   R = Alkyl (Phenyl) R_2N = Piperidino,
                                                        Morpholino

Enamine aromatischer Methylketone (z.B. *331*) überführt Schwefelkohlenstoff ebenfalls in guten Ausbeuten in 2.6-Diaryl-α-thiopyrone *334*.

Da diese Enamine sich nicht mit sich selbst kondensieren, müssen *332* und *333* — die auch für das vorangehende Beispiel nicht auszuschließen sind — als Zwischenstufen postuliert werden [156]. Plausibler erscheint der Ringschluß von *331* und *332* zu *333a*, das glatt in *334* übergehen sollte.

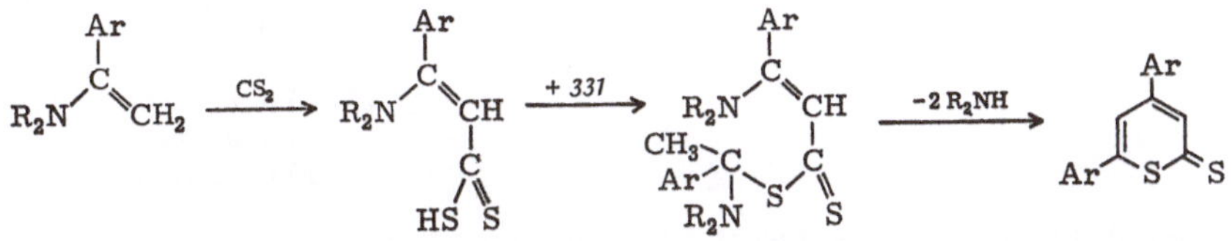

331   *332*   *333*   *334*

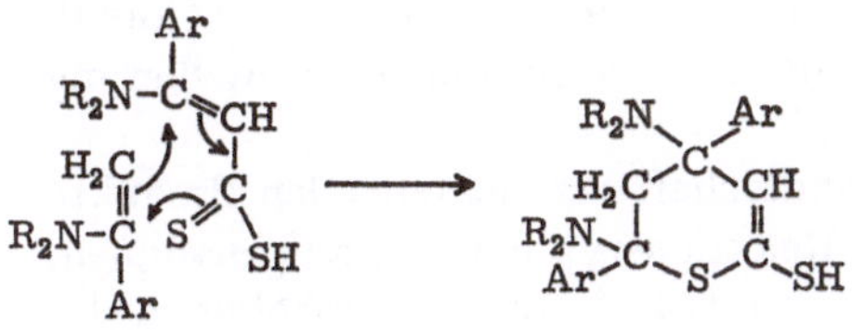

331   *332*   *333a*

Von Ketonen abgeleitete Schiffsche Basen, z.B. *335*, die mit der Enaminform im Gleichgewicht stehen, addieren Schwefelkohlenstoff zu isolierbaren Dithiosäuren *336*.

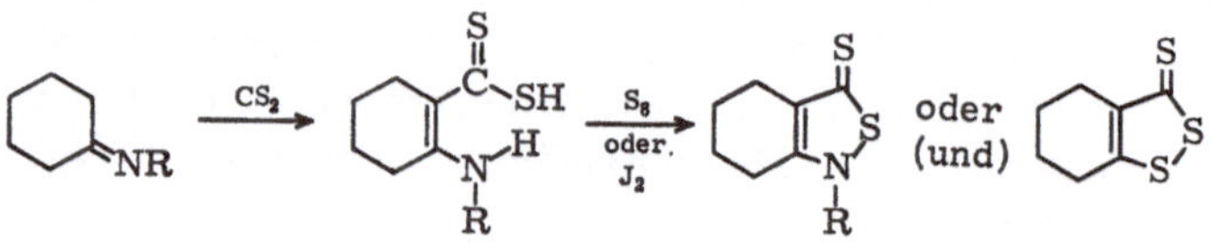
NR
CS_2
S
C
SH
H
N
R
S_8
oder
J_2
S
S
N
R
oder
(und)
S
S
S
335
336
337
338

Diese schließen oxidativ den Ring zu Isothiazolderivaten *337*. Verwendet man Schwefel, so entstehen unter S-Thiolierung und Aminabspaltung auch 1.2-Dithiol-thione-3 *338*, in DMF bei höherer Temperatur häufig als Hauptprodukt [156]. Mit „normalen" Ketonenaminen wie z.B. *93* sind bei —30 °C in Acetonitril/Äther instabile Zwitterionen *339* nachzuweisen, die bei der Umsetzung mit Schwefel in Acetonitril bei —30 °C Verbindung *338*, bei der Umsetzung mit Phenyliso- und -thiocyanat *340a—c* ergeben [157].

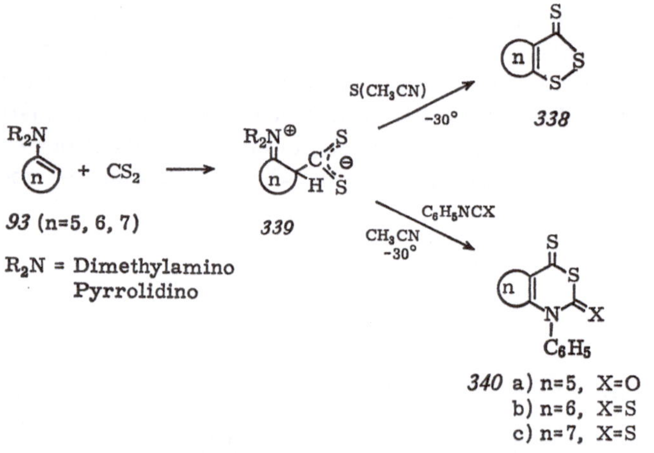
R_2N
n
+ CS_2
93 (n=5, 6, 7)
R_2N = Dimethylamino
Pyrrolidino
$R_2N^{\oplus}$
S
C
⊖
H
S
n
339
$S(CH_3CN)$
-30°
S
n
S
S
338
C_6H_5NCX
CH_3CN
-30°
S
n
S
N
X
C_6H_5
340 a) n=5, X=O
b) n=6, X=S
c) n=7, X=S

3.6. Acylierung in Gegenwart von Schwefel

Mit diesem Reaktionstyp wird das bisher eingehaltene Thema überschritten, da in allen Fällen primär der Schwefel das Enamin elektrophil, d.h. oxidierend in β-Stellung angreift, und die nachfolgende Acylierung nunmehr am Schwefel einsetzt [158].

Wegen der strukturellen Verwandtschaft der entstehenden Produkte mit den besprochenen, wird dieser Reaktionstyp hier mit aufgenommen. Die komplizierten Verhältnisse seien am Beispiel des 1-Morpholino-cyclohexen-1 (*341*) erläutert [156].

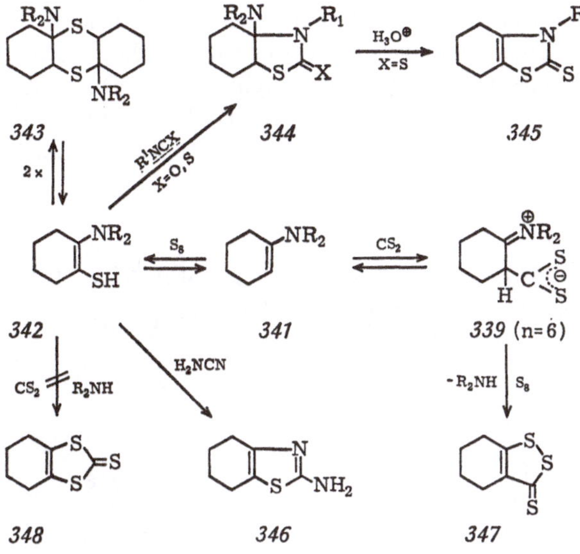

$$343 \qquad 344 \qquad 345$$

$$342 \qquad 341 \qquad 339 \ (n=6)$$

$$348 \qquad 346 \qquad 347$$

R_2N = Morpholino

Die Reaktion mit Schwefel in einem polaren Lösungsmittel führt zunächst zum Thiol *342*, das sich unter veränderten Bedingungen als Dimeres *343* fassen läßt. Die Existenz von *342* geht zusätzlich aus dem glatten Ringschluß mit Isocyanaten [159)] und Isothiocyanaten zu *344* hervor. *344* (X = S) wird von Säuren in Thiazolderivate *345* übergeführt. Schließlich reagiert Cyanamid mit *342* nur dann zum Aminothiazol *346*, wenn *gleichzeitig* Schwefel anwesend ist. Die Reaktion zu *342* muß rückläufig sein, denn auf Zusatz von Schwefelkohlenstoff entsteht nur das 1.2-Dithiol-3-thion *347* und — genau wie mit anderen alkylsubstituierten Enaminen — nicht das 1.3-Dithiol-thion-2 *348*. Der Ringschluß zu *347* ist aber nicht über *342* sondern nur über *339* (*n* = 6) möglich.

Verfügt das Enamin über Arylgruppen, so tritt auch der Heterocyclus *348* im Gemisch mit auf, im Falle von *349* bis 50%, im Falle von *351* entsteht ausschließlich *352* [156)].

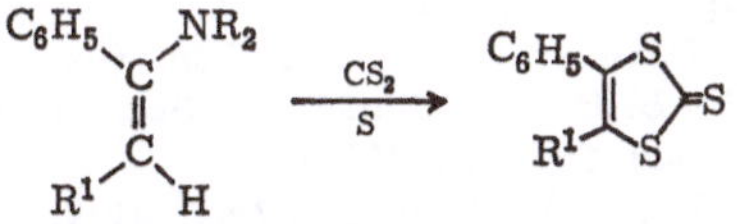

$$349: R^1 = C_6H_5 \qquad 350$$
$$351: R^1 = C_6H_{13} \qquad 352$$

1.3-Dithiol-thione-2 mit aliphatischen Substituenten sind jedoch ebenfalls aus Enaminen zugänglich, wenn die oxidative Einführung des Schwefels mittels Thiuramdisulfids geschieht [160]. Die postulierte Zwischenstufe *353*, deren Struktur durch die Hydrolyse zum bekannten *354* gestützt wird, setzt sich mit Schwefelwasserstoff zu *348* um.

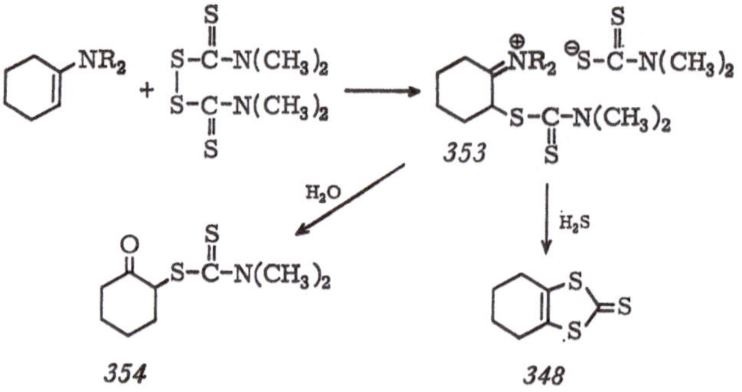

Die Ausbeuten liegen hier und in ähnlichen Fällen zwischen 50—70%.

4. Literatur

[1] *Coenen, M.:* Angew. Chem. *61*, 11 (1949).

[2] *Stork, G., R. Terrell,* and *J. Szmuszkovicz:* J. Am. Chem. Soc. *76*, 2029 (1954).

[3] —, and *H. K. Landesman:* J. Am. Chem. Soc. *78*, 5128 (1956).

[4] vgl. die Zusammenfassungen:

[a] *Szmuszkovicz, J.:* Advances in Organic Chemistry, Vol. 4, pp. 1—113. New York: Interscience Publishers 1963;

[b] *Bláha, K.,* and *O. Červinka:* Advan. Heterocyclic Chem., Vol. 16, p. 147. New York–London: Academic Press 1966;

[c] *de Poorter, H.:* Mededelingen van de Vlaamse, S. 100. Chemische Vereinigung 1966;

[d] *Hochrainer, A.:* Oesterr. Chemiker-Ztg. *66*, 355 (1965);

[e] *Cook, A. G.:* Enamines: Synthesis, Structure and Reactions. New York–London: M. Dekker 1969.

[5] vgl. Zusammenfassung *Opitz, G.:* Angew. Chem. *79*, 161 (1967).

[6] *Borrmann, D.:* Methoden der organ. Chemie (Houben-Weyl), 4. Aufl., Bd. VII/4, S. 53. Stuttgart: Georg Thieme 1968.

[7] *Mannich, C.,* u. *H. Davidsen:* Ber. Deut. Chem. Ges. *69*, 2106 (1936).

[8] *Heyl, F. W.,* and *M. E. Herr:* J. Am. Chem. Soc. *75*, 1918 (1953).

[9] *Hoch, J.:* Compt. Rend. *200*, 938 (1935).

[10] *Nelson, P.,* and *A. Pelter:* J. Chem. Soc. *1965*, 5142.

[11] *v. Hirsch, H.:* Chem. Ber. *100*, 1289 (1967).

[12] [a] *White, W. A.,* and *H. Weingarten:* J. Org. Chem. *32*, 213 (1967);

[b] *Hauthal, H. G.,* u. *D. Scheid:* Z. Chem. *9*, 62 (1969).

[13] *Leonard, N. J., A. S. Hay, R. W. Fulmer,* and *V. W. Gash:* J. Am. Chem. Soc. 77, 439 (1955) und l.c. 4[a].

[14] *Stamhuis, E. J., W. Maas,* and *H. Wynberg:* J. Org. Chem. 30, 2160 (1965). Vgl. E. J. Stamhuis sowie G. H. Alt in Lit. 4[e].

[15] *Opitz, G.:* Liebigs Ann. Chem. 650, 122 (1961). — *Brannock, K. C.,* and *R. D. Burpitt:* J. Org. Chem. 26, 3576 (1961). — *Opitz, G.,* u. *H. Mildenberger:* Liebigs Ann. Chem. 649, 26 (1961).— *Opitz, G., H. Mildenberger* u. *H. Suhr:* Liebigs Ann. Chem. 649, 47 (1961) s. auch Lit. [2,3,106,107].

[16] *Otto, P., L. A. Feiler* u. *R. Huisgen:* Angew. Chem. 80, 759 (1968).

[17] *Huisgen, R., L. A. Feiler,* and *P. Otto:* Tetrahedron Letters 1968, 4486.

[18] *Stamhuis, E. J.,* and *W. Maas:* J. Org. Chem. 30, 2156 (1965). Vgl. E. J. Stamhuis sowie G. H. Alt in Lit. 4[e].

[19] *Hoch, H.:* Dissertation Universität Würzburg 1968, Chem. Ber. in Vorbereitung. Weitere Enamine vgl. S. K. Malhotra in Lit. 4[e] sowie K. Nagarajan und S. Rajappa, Tetrahedron Letters 1969, 2293.

[20] [a] *Gurowitz, W. D.,* and *M. A. Joseph:* J. Org. Chem. 32, 3289 (1967); [b] — — Tetrahedron Letters 1965, 4433; [c] *Nagarajan, K.,* and *S. Rajappa:* Tetrahedron Letters 1969, 2293.

[21] *Opitz, G.,* u. *A. Griesinger:* Liebigs Ann. Chem. 665, 101 (1963). — *Opitz, G., A. Griesinger* u. *H. W. Schubert:* Liebigs Ann. Chem. 665, 91 (1963).

[22] —, *H. Hellmann* u. *H. W. Schubert:* Liebigs Ann. Chem. 623, 117 (1959). Vgl. S. K. Malhotra sowie J. V. Paukstelis in Lit. 4[e].

[23] [a] *Castille, A.,* et *E. Ruppol:* Bull. Soc. Chim. Biol. 10, 623 (1928). — *Truce, W. E.,* and *J. A. Simms:* J. Org. Chem. 22, 762 (1957); [b] s.l.c. [13].

[24] [a] *Leonard, N. J.,* and *V. W. Gash:* J. Am. Chem. Soc. 76, 2781 (1954); s. auch l.c. [22] und dort unter l.c. [1] angegebene Literaturzitate; [b] *Prelog, V.,* u. *O. Häfliger:* Helv. Chim. Acta 32, 1851 (1949).

[25] *Tschesche, R.,* u. *G. Snatzke:* Chem. Ber. 90, 579 (1957).

[26] *Grob, C. A., A. Kaiser* u. *E. Renk:* Helv. Chim. Acta 40, 2170 (1957).

[27] *Cook, A. G.:* J. Am. Chem. Soc. 85, 648 (1963).

[28] *Leonard, N. J.,* and *A. G. Cook:* J. Am. Chem. Soc. 81, 5627 (1959).

[29] —, and *W. K. Musker:* J. Am. Chem. Soc. 81, 5631 (1959).

[30] *Hünig, S., E. Benzing* u. *E. Lücke:* Chem. Ber. 90, 2833 (1957).

[31] *Schultes, H.:* Angew. Chem. 47, 258 (1934); s. auch Lit. [30].

[32] *Sauer, J.,* and *H. Prahl:* Tetrahedron Letters 1966, 2863; Chem. Ber. 102, 1917 (1969).

[33] *Leonard, N. J., R. W. Fulmer,* and *A. S. Hay:* J. Am. Chem. Soc. 78, 3457 (1956).

[34] [a] *Fusco, R., G. Bianchetti* e *S. Rossi,* Gazz. Chim. Ital. 91, 825 (1961); [b] *Bianchetti, G., D. Pocar, P. Dalla Croce* u. *A. Vigevani:* Chem. Ber. 98, 2715 (1965); [c] —, *P. Ferruti* e *D. Pocar:* Gazz. Chim. Ital. 97, 579 (1967); [d] *Pocar, D., G. Bianchetti* e *P. Ferruti:* Gazz. Chim. Ital. 97, 597 (1967); [e] *Jaquier, R., C. Petrus* et *F. Petrus:* Bull. Soc. Chim. France 1966, 2845; [f] *Pocar, D., R. Stradi* e *B. Gioia:* Gazz. Chim. Ital. 98, 958 (1968).

[35] *Hünig, S., K. Hübner* u. *E. Benzing:* Chem. Ber. 95, 926 (1963).

[36] *Berchtold, G. A.:* J. Org. Chem. 26, 3043 (1961).

[37] *Mannich, C.,* u. *E. Kniss:* Ber. Deut. Chem. Ges. 74, 1629, 1637 (1941).

[38] *Stork, G., A. Brizzolara, H. Landesman, J. Szmuszkovicz,* and *R. Terrell:* J. Am. Chem. Soc. 85, 207 (1963).

[39] *Bianchetti, G., P. Dalla Croce* e *D. Pocar:* Rend. Ist. Lomb. 99, 259 (1966).

S. Hünig und H. Hoch

40) s. auch *Hellmann, H.*, u. *G. Opitz*: α-Aminoalkylierung, S. 188. Weinheim/Bergstr.: Verlag Chemie 1966.

41) *Leonard, N. J.*, and *F. P. Hauck jr.*: J. Am. Chem. Soc. 79, 5279 (1957).

42) *Schöpf, C., R. Klug* u. *R. Rausch*: Liebigs Ann. Chem. 616, 151 (1958).

43) *Larive, H.*, u. *R. Dennilauer*: Chimia 15, 115 (1961) und spätere Arbeiten von *Metzger, J., H. Larivé, J. Vincent* et *R. Dennilauer*: J. Chim. Phys. 60, 467 (1963);
   a) Nur bei langsamer Zugabe des 2,3-Dimethylbenzthiazoliumsalzes zu überschüssiger Natronlauge ist das monomere Enamin isolierbar. J. R. Owen THL 1969, 2709.

44) *Kröhnke, F., H. Dickhäuser* u. *I. Vogt*: Liebigs Ann. Chem. 644, 93 (1961).

45) DRP 154448/20. 9. 1903 Farbwerke Hoechst AG.
   a) Vgl. *A. G. Cook* sowie *M. E. Kuehne* in Lit. 4e).

46) *Hasek, R. H.*, and *J. C. Martin*: J. Org. Chem. 28, 1468 (1963).

47) *Martin, J. C., P. G. Gott*, and *H. U. Hostettler*: J. Org. Chem. 32, 1654 (1967).

48) *Otto, P., L. A. Feiler* u. *R. Huisgen*: Angew. Chem. 80, 759 (1968).

49) *Huisgen, R.*, and *P. Otto*: J. Am. Chem. Soc. 91, 5922 (1969).

50) *Gleiter, R.*, u. *R. Hoffmann*: Angew. Chem. 81, 225 (1969).

51) *Gompper, R.*: Angew. Chem. 81, 348 (1969).

52) *Wagner, H. U.*, u. *R. Gompper*: unveröffentlichte Ergebnisse.

53) *Kuehne, M. E.*, and *P. J. Sheevan*: J. Org. Chem. 33, 4406 (1969).

54) s. Zusammenfassung in *Ulrich, H.*: Cycloaddition Reactions of Heterocumulenes, Vol. 9 der Serie Organic Chemistry, S. 38 ff u. l. c. 48). New York–London: Academic Press 1967.

55) a) *Hasek, R. H., P. G. Gott*, and *J. C. Martin*: J. Org. Chem. 31, 1931 (1966) sowie Zusammenfassung in l. c. 54);
   b) unveröffentlichte Versuche von *H. Hoch*, Universität Würzburg 1968–1969.

56) *Opitz, G., H. Adolph, M. Kleemann* u. *F. Zimmermann*: Angew. Chem. 73, 654 (1961).

57) —, u. *M. Kleemann*: Liebigs Ann. Chem. 665, 114 (1963).

58) *Berchtold, G. A.*: J. Org. Chem. 26, 4776 (1961) und l.c. 55).

59) *Hoch, H.*: Chem. Ber., in Vorbereitung.

60) *Hünig, S.*, and *H. Hoch*: Tetrahedron Letters 1966, 5215. — *Kirrmann, A.*, et *C. Wakselman*: Bull. Soc. Chim. France 1967, 3766.

61) —, *H. J. Buysch, H. Hoch* u. *W. Lendle*: Chem. Ber. 100, 3996 (1967).

62) *Opitz, G.*, u. *F. Zimmermann*: Liebigs Ann. Chem. 662, 178 (1963).

63) *Criegee, R., G. Bolz* u. *R. Askani*: Chem. Ber. 102, 275 (1969).

64) *Reichardt, Chr.*, u. *K. Dimroth*: Fortschr. Chem. Forsch. 11, 1 (1968).

65) *Grob, C. A.*, u. *H. J. Wilkens*: Helv. Chim. Acta 50, 725 (1967);
   a) *Kashima, Ch., M. Yamamoto*, and *N. Sugiyama*: J. Chem. Soc. (C) 1970, 111;
   b) *Leonard, N. J.*, and *J. A. Adamcik*: J. Am. Chem. Soc. 81, 596 (1959).

66) *Dabrowsky, J.*, u. *U. Dabrowska*: Chem. Ber. 101, 3392 (1968). — *Kaminska-Frela, K.*: Spektrochim. Acta 1966, 211;
   a) *Meyers, A. J., A. H. Reine, J. C. Sircar, K. B. Rao, S. Singh, H. Weidmann*, and *M. Fitzpatrick*: J. Heterocyclic Chem. 5, 151 (1968).

67) *Jacquier, R.*, et *G. Maury*: Bull. Soc. Chim. France 1967, 320.

68) *Opitz, G.*, u. *E. Tempel*: Liebigs Ann. Chem. 699, 74 (1966).

69) *Berchtold, G. A., R. R. Harvey*, and *G. E. Wilson jr.*: J. Org. Chem. 30, 2642 (1965).

70) *Staudinger, H.*, u. *R. Endle*: Liebigs Ann. Chem. 401, 263 (1913).

71) *Hopff, H.*, and *W. Rapp*: U.S.A. Pat. 2,265,165; C. A. 36, 1614 (1942).

72) Nur eine weitere C-Acylierung ist bekannt vgl. Abschn. 2.5 (197).

73) *Hünig, S.,* u. *E. Lücke:* Chem. Ber. *92,* 652 (1959).
    a) Vgl. *M. E. Kuehne* in Lit. [4e].
74) *Eistert, B.,* u. *R. Wessendorf:* Chem. Ber. *94,* 2590 (1961).
75) *Hünig, S.,* u. *W. Lendle:* Chem. Ber. *93,* 909, 913 (1960).
76) s. zusammenfassender Artikel über Erzeugung und Reaktionen von Sulfen in l.c. [5] und dort zitierte Literatur.
77) *Gandini, A., P. Schenone* u. *B. Bignardi:* Monatsh. Chem. *98,* 1518 (1967).
78) *Lücke, E.:* Dissertation, Universität Marburg 1958 und l.c. [30].
79) *Hünig, S., E. Lücke* u. *E. Benzing:* Chem. Ber. *91,* 129 (1958).
80) —, u. *M. Salzwedel:* Chem. Ber. *99,* 823 (1966).
81) —, u. *W. Eckardt:* Chem. Ber. *95,* 2493 (1962).
82) *Hamrick jr., P. J., C. F. Hauser,* and *C. R. Hauser:* J. org. Chem. *24,* 583 (1959); dort weitere Literaturhinweise.
83) *Hünig, S.,* u. *M. Salzwedel:* Angew. Chem. *71,* 339 (1959).
84) —, *E. Lücke,* and *W. Brenninger:* Organic Synthesis, Vol. *43,* S. 34. New York: John Wiley & Sons 1963.
85) *Lendle, W.:* Dissertation, Universität Marburg 1959.
86) *Hünig, S.,* u. *H. J. Buysch:* Chem. Ber. *100,* 4010 (1967).
87) — — Chem. Ber. *100,* 4017 (1967).
88) *Borrmann, D.:* Methoden der organ. Chemie (Houben-Weyl), 4. Aufl., Bd. VII/4, S. 226. Stuttgart: Georg Thieme Verlag 1968.
89) *Martin, J. C., R. D. Burpitt,* and *H. U. Hostettler:* J. Org. Chem. *32,* 210 (1967).
90) *Hünig, S., E. Benzing* u. *K. Hübner:* Chem. Ber. *94,* 486 (1961); s. auch *Millward, B. B.:* J. Chem. Soc. *1960,* 26 und l.c. [60].
91) *Rao, C. N. R.:* Chemical Applications of Infrared Spectroscopy, p. 233. New York–London: Academic Press 1963.
92) *Scott, A. I.:* Interpretation of the UV-Spectra of natural products, p. 140. Pergamon Press 1964.
93) *Hanford, W. E.,* and *J. C. Sauer:* Org. Reactions Vol. III, Chap. 3, p. 108. New York: John Wiley & Sons 1946.
94) *Alt, G. H.:* J. Org. Chem. *31,* 2384 (1966).
95) für $n = 6$: *Helmers, R.:* Diplomarbeit, Universität Marburg 1961.
96) für $n = 5$: *Campbell, R. D.,* and *W. L. Harmer:* J. Org. Chem. *28,* 379 (1963).
97) *Helmers, R.:* Acta Chem. Scand. *19,* 2139 (1965).
98) — Tetrahedron Letters *1966,* 1905.
99) *Campbell, R. D.,* and *J. A. Jung:* J. Org. Chem. *30,* 3711 (1965).
100) *Walborsky, H. M.:* J. Am. Chem. Soc. *74,* 4962 (1952); s. auch *Erickson, J. L.E., F. E. Collins jr.,* and *B. L. Owen:* J. Org. Chem. *31,* 480 (1966).
101) *Mühlstaedt, M.,* u. *J. Reimer:* Z. Chem. *4,* 70 (1964).
102) *Alt, G. H.,* and *A. J. Speziale:* J. Org. Chem. *29,* 798 (1964).
103) — — J. Org. Chem. *30,* 1407 (1965).
104) *Inukai, T.,* and *R. Yoshizawa:* J. Org. Chem. *32,* 404 (1967).
105) s.l.c. [38] und *Fleming, I.,* and *J. Harley-Mason:* J. Chem. Soc. *1964,* 2165. — *Westman, T. L., R. Paredes,* and *W. S. Brey jr.:* J. Org. Chem. *28,* 3512 (1963). — *Westman, T. L.,* and *A. E. Kober:* J. Org. Chem. *29,* 2448 (1964). — *Augustine, R. L.,* and *H. V. Cortez:* Chem. Ind. 490 (*1963*). — *House, H. O., M. B. Trost, R. W. Magin, R. G. Carlson, R. W. Franck,* and *G. H. Rasmusson:* J. Org. Chem. *30,* 2513 (1965).
106) *Hickmott, P. W.,* and *B. J. Hopkins:* J. Chem. Soc. *1968,* 2918.
107) *Hargreaves, J. R., P. W. Hickmott,* and *B. J. Hopkins:* J. Chem. Soc. 1968, 2599.
108) — — — J. Chem. Soc. *1969,* 592.

S. Hünig und H. Hoch

109) *Hickmott, P. W.,* and *J. R. Hargreaves:* Tetrahedron *23,* 3151 (1967).
110) a) *Fischer, G. W.,* u. *W. Schroth:* Chem. Ber. *102,* 590 (1969);
     b) — — Chem. Ber. *102,* 575 (1969).
111) vgl. zu dieser Gegenüberstellung l.c. ³⁰⁾ und *Jacquier, R.,* et *G. Maury:* Bull. Soc. Chim. France *1967,* 316.
112) *Boyd, G. V.,* and *D. Hewson:* Chem. Commun. *1965,* 536. — *Boyd, G. V., D. Hewson,* and *R. A. Newberry:* J. Chem. Soc. *1965,* 935.
     a) Zur Konfiguration solcher Cyanine vgl. *K. Feldmann, E. Daltrozzo* und *G. Scheibe,* Z. Naturfosch. *22b,* 722 (1967).
113) *Risaliti, A.:* Boll. Sci. Fac. Chim. Ind. Bologna *19,* 173 (1961) (C. A. *58,* 4390f (1963)).
114) *Ziegenbein, W.:* Angew. Chem. *77,* 380 (1965).
115) *Ito, Y., S. Katsuragawa, M. Okano,* and *R. Oda:* Tetrahedron *23,* 2159 (1967).
116) vgl. z.B. *Eilingsfeld, H., M. Seefelder* u. *H. Weidinger:* Chem. Ber. *96,* 2899 (1963).
117) *Oxley, F.,* and *W. F. Short:* J. Chem. Soc. *1948,* 1514.
118) *Meuer, V.:* Dissertation, Universität Würzburg 1962.
119) *Brunings, K. J.,* and *A. H. Corwin:* J. Am. Chem. Soc. *64,* 593 (1942).
120) *Brooker, L. G. S., F. L. White, R. H. Sprague, S. G. Dent jr.,* and *G. van Zandt:* Chem. Rev. *41,* 325 (1947).
121) *Hünig, S., W. Gräßmann, V. Meuer* u. *E. Lücke:* Chem. Ber. *100,* 3024 (1967).
122) — — — — u. *W. Brenninger:* Chem. Ber. *100,* 3039 (1967).
123) *Bianchetti, G., D. Pocar* e *P. Dalla Croce:* Gazz. Chim. Ital. *93,* 174 (1963).
124) *Fusco, R., G. Bianchetti* e *D. Pocar:* Gazz. Chim. Ital. *91,* 1233 (1961).— *Pocar, D., S. Maiorana* e *P. Dalla Croce:* Gazz. Chim. Ital. *98,* 949 (1968).
125) *Munk, M. E.,* and *Yung Ki Kim:* J. Am. Chem. Soc. *86,* 2213 (1964). — *Huisgen, R., M. Seidel, G. Wallbillich,* and *H. Knupfer:* Tetrahedron *17,* 3 (1962).
126) *Nikolajewski, H. E., S. Dähne, D. Leupold,* and *B. Hirsch:* Tetrahedron *24,* 6685 (1968).
127) *Sobotka, W., W. N. Beverung, G. G. Munoz, J. C. Sircar,* and *A. I. Meyers:* J. Org. Chem. *30,* 3667 (1965). — *Meyers, A. I.,* and *J. C. Sircar:* J. Org. Chem. *32,* 1250 (1967). — *Meyers, A. I., A. H. Reine,* and *R. Gault:* J. Org. Chem. *34,* 698 (1969).
127) a) vgl. auch *Bohlmann, F.,* u. *O. Schmidt:* Chem. Ber. *97,* 1354 (1964).
128) *Möller, F.:* Methoden der organ. Chemie (Houben-Weyl), 4. Aufl., Bd. XI/1, S. 985. Stuttgart: Georg Thieme Verlag 1957.
129) *Hünig, S.,* u. *M. Kiessel:* J. Prakt. Chem. 224 (1958).
130) *Wolinsky, J.,* and *D. Chan:* Chem. Commun. *1966,* 567;
     a) *Elkik, E.,* et *P. Vandescal:* Compt. Rend. *1967,* 1779.
131) *Groß, H.,* u. *J. Rusche:* Angew. Chem. *76,* 534 (1964).
132) Dtsch. Bundes-Pat. 27.5.57 Farbenfabriken Bayer AG, Leverkusen.
133) *Kuehne, M. E.:* J. Am. Chem. Soc. *81,* 5400 (1959).
134) *Grigat, E., R. Pütter* u. *E. Mühlbauer:* Chem. Ber. *98,* 3777 (1965).
135) *Martin, D.,* u. *S. Rackow:* Chem. Ber. *98,* 3662 (1965).
136) *Fusco, R., S. Rossi* e *G. Bianchetti:* Gazz. Chim. Ital. *91,* 481 (1961).
137) *Chrétien-Bessière, Y.,* et *H. Sesne:* Compt. Rend. *266,* 635 (1968).
138) a) *Hünig, S., K. Hübner* u. *E. Benzing:* Chem. Ber. *95,* 926 (1962);
     b) *Petersen, S.:* Methoden der org. Chemie (Houben-Weyl), 4. Aufl., Bd. VIII, S. 129. Stuttgart: Georg Thieme 1952;
     c) *Naegeli, C., A. Tyabji* u. *L. Conrad:* Helv. Chim. Acta *21,* 1127 (1938).
139) *Bestian, H.:* Angew. Chem. *80,* 304 (1968). — *Graf, R.:* Liebigs Ann. Chem. *661,* 111 (1963).

140) *Browne, D. W.*, and *G. N. Dyson:* J. Chem. Soc. *1931*, 3285; s. auch l.c. [138a)].

141) *Perelman, M.*, and *S. A. Mizsak:* J. Am. Chem. Soc. *84*, 4988 (1962).

142) *Opitz, G.*, and *J. Koch:* Angew. Chem. Intern. Ed. Engl. *2*, 152 (1963).

143) *Bose, A. K.*, and *G. Mina:* J. Org. Chem. *30*, 812 (1965).

144) *Ley, K., U. Eholzer*, and *R. Nast:* Angew. Chem. Intern. Ed. Engl. *4*, 519 (1965).

145) a) s.l.c. [138a)] und l.c. [19)];
   b) *Bianchetti, G., P. Dalla Croce* e *D. Pocar:* Gazz. Chim. Ital. *94*, 606 (1964);
   c) *Fusco, R., G. Bianchetti* e *S. Rossi:* Gazz. Chim. Ital. *91*, 825 (1961);
   d) *Ried, W.*, und *W. Käppeler:* Liebigs Ann. Chem. *673*, 132 (1964);
   e) *Clemens, D. H.*, and *W. D. Emmons:* J. Org. Chem. *26*, 767 (1961).

146) a) *Berchtold, G. A.:* J. Org. Chem. *26*, 3043 (1961);
   b) *Matterstock, U.*, u. *H. Jensen:* Ger. Pat. 1168896 (1964).

147) *Schneider, W., K. Gaertner* u. *A. Jordan:* Ber. Deut. Chem. Ges. *57*, 523 (1924).
   — *Schneider, W.:* Liebigs Ann. Chem. *438*, 132 (1924).

148) *Mumm, O., H. Hinz* u. *J. Diederichsen:* Ber. Deut. Chem. Ges. *72*, 2107 (1939).
   — *Clifford, A. F.*, and *C. S. Kobayaski:* J. Inorg. Chem. *4*, 571 (1965).

149) *Hünig, S.*, u. *K. Hübner:* Chem. Ber. *95*, 937 (1962).

150) *Blatter, H. M.*, and *H. Lukaszewski:* J. Org. Chem. *31*, 722 (1966).

151) vgl. auch *de Stevens, G., B. Smolinsky, J. Wojtkunski* u. *R. W. J. Carney:* Angew. Chem. *76*, 194 (1964). — *de Stevens, G., B. Smolinsky* and *L. Dorfman:* J. Org. Chemistry *29*, 1115 (1964).

152) *Bianchetti, G., D. Pocar* e *S. Rossi:* Gazz. Chim. Ital. *93*, 255 (1963).

153) s. auch l.c. [145e)].

154) *Goerdeler, J.*, u. *U. Keuser:* Chem. Ber. *97*, 2209 (1964).

155) s. auch l.c. [148)].

156) s. Review-Artikel *Mayer, R.*, u. *K. Gewald:* Angew. Chem. *79*, 298 (1967) und dort zitierte Literatur.

157) *Gompper, R., B. Witzel*, and *W. Elser:* Tetrahedron Letters *1968*, 5519.

158) vgl. den oxidativen Angriff mit Benzolperoxyd (*R. L. Augustine*): J. Org. Chem. *28*, 581 (1963);
   mit Thalliumtriazetat (*M. E. Kuehne* und *T. J. Giacobbe*): J. Org. Chem. *33*, 3359 (1968);
   mit Diarylsulfiden (*M. E. Kuehne*): J. Org. Chem. *28*, 2124 (1963).

159) *Ley, K.*, u. *R. Nast:* Angew. Chem. *77*, 544 (1965).

160) *Fanghänel, E.:* Z. Chem. *74*, 70 (1964).

Eingegangen am 1. August 1969

Nitrogen Ylids

W. K. Musker

Associate Professor of Chemistry, University of California, Davis, CA 95616

Contents

I. Introduction

In general, quaternary ammonium ions may be expected to react with base at three different positions [58]: a) at an α-carbon to affect a nucleophilic displacement reaction which yields the free amine and alkylated base;

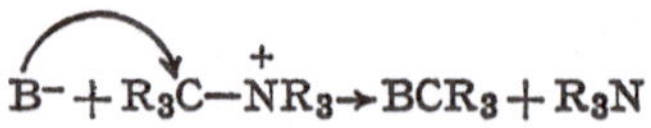

$$B^- + R_3C-\overset{+}{N}R_3 \rightarrow BCR_3 + R_3N$$

b) at a β-hydrogen to affect a *Hofmann* or β-elimination reaction which yields alkene, amine, and the conjugate acid of the attacking base;

$$B^- + R_2C-CH_2-\overset{+}{N}R_3 \rightarrow BH + R_2C=CH_2 + R_3N$$

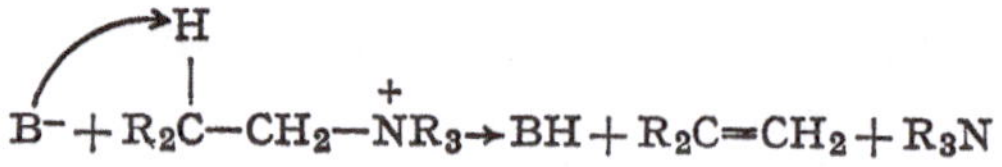

c) at an α-hydrogen which yields the conjugate acid of the attacking base and a dipolar carbanionic species or nitrogen ylid.

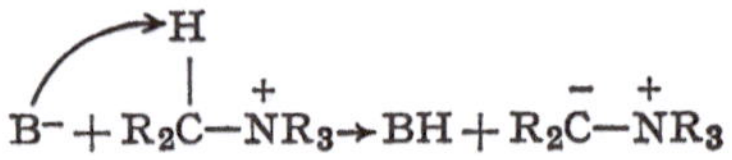

$$B^- + R_2C-\overset{+}{N}R_3 \rightarrow BH + R_2\overset{-}{C}-\overset{+}{N}R_3$$

The name "ylid" originated in *Wittig*'s writings [161–162] when he called the species prepared by attack of strong base on the tetramethylammonium ion, trimethylammoniummethylide *1*. This species, or its alkali metal derivative, will be referred to as "the ylid" in this discussion.

$$(CH_3)_3\overset{+}{N}-\overset{-}{C}H_2$$

"the ylid"

1

In this review the conditions under which α-proton abstraction occurs are examined and the properties and reactions of the resulting nitrogen ylid are described. The subject of nitrogen ylids was reviewed by *Johnson* [78] in 1966 in a chapter in the book entitled "Ylid Chemistry". In

Johnson's discussion [78] a major emphasis is placed on those nitrogen ylids in which either the positive or negative charge can be delocalized by interaction with an unsaturated system. In this review the main emphasis is placed on *nitrogen ylids where no delocalization is possible:* the ylid resulting from the deprotonation of the tetramethylammonium cation will receive major attention although the reactions of cations containing an N^+-CH_3 group will also be discussed. A brief summary of the reactions of the tetramethylammonium ion was published in 1968 [108].

Historically, the development of the chemistry of nitrogen ylids began with early attempts to prepare organic derivatives containing pentavalent nitrogen. To this purpose, *Schlenk* and *Holtz* [126] treated tetramethylammonium chloride with sodium triphenylmethide and with benzylsodium, isolating tetramethylammonium triphenylmethide and tetramethylammonium benzylide, respectively. These compounds were thought at the time to be pentavalent nitrogen compounds.

$$(C_6H_5)_3\overset{-+}{C}Na + (CH_3)_4\overset{+}{N}\overset{-}{Cl} \rightarrow (CH_3)_4\overset{+}{N}\overset{-}{C}(C_6H_5)_3 + Na^+ + Cl^-$$

Later *Hager* and *Marvel* [56] attempted to prepare analogous compounds in which all five groups around nitrogen were more equivalent. These workers found that the reaction of triethylbenzylammonium bromide with ethyllithium did not produce tetraethylammonium benzylide, ruling out the existence of any intermediate in which the five groups bound to nitrogen approached equivalency. From this observation *Hager* and *Marvel* concluded that the materials prepared by *Schlenk* and *Holtz* were tetraalkylammonium salts of the relatively stable triphenylmethyl and benzyl carbanions rather than derivatives of pentavalent nitrogen. Since no toluene or triphenylmethane was found, the benzyl and triphenylmethyl carbanions apparently are not strong enough bases to effect proton abstraction from the tetramethylammonium ion under the conditions of this reaction. In addition, no products resulting from displacement reactions were reported [126].

In 1944 *Wittig* and *Felletschin* [161] began a reinvestigation of the pentavalent nitrogen problem and succeeded in isolating a red powder from the treatment of a 9-fluorenyltrimethylammonium bromide, *2*, with phenyllithium in ether. However, since benzene was isolated from the reaction mixture, the compound could not be a pentavalent nitrogen derivative and was assigned an ylid structure on the basis of its reactions with water, methyl iodide, iodine, and benzyl bromide.

Following this initial preparation of a stable material having an ylid structure, a variety of phosphorus, arsenic, and sulfur ylids have been prepared and characterized and their chemistry has been thoroughly reviewed [78]. The chemistry of trimethylamine imine [2] and trimethylamine oxide, compounds which are isoelectronic with the ylid, have been reviewed [30,78] and will not be described here.

From this beginning, the chemistry of numerous nitrogen ylids has been studied, but the ylid, the simplest member of this series, has been studied most extensively and will be discussed first.

II. Reactions of the Tetramethylammonium Ion with Organolithium Reagents. — "The Ylid"

A. Preparation and Structure

In 1947 *Wittig* and *Wetterling* [162] examined the reaction between tetramethylammonium bromide and phenyllithium in ether in an effort to prepare tetramethylammonium phenyl. None of the desired product was obtained, but benzene was formed along with an insoluble material which was characterized as the lithium bromide complex of the ylid on the basis of its reactions with water, iodine, methyl iodide, and benzophenone.

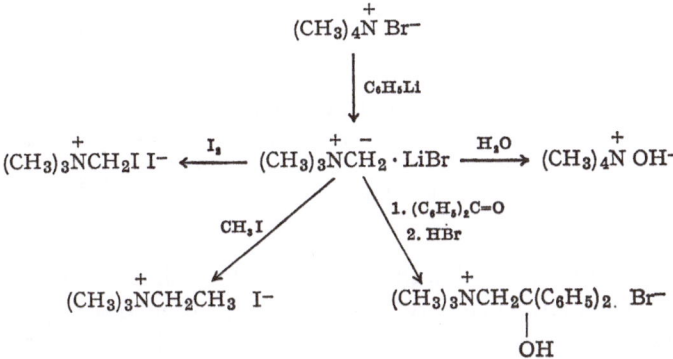

The same material is also obtained when bromomethyltrimethyl-ammonium bromide is treated with phenyllithium in ether. Although the lithium bromide complex of the ylid is insoluble in ether, the solid can be dissolved in tetrahydrofuran (THF) in concentrations up to 0.9 M [39].

The actual structure of this complex, however, was not completely clear, and additional work [170] was done in an effort to determine whether the material should be formulated as a lithium bromide complex of the dipolar ylid or lithiomethyltrimethylammonium bromide. Because of this uncertainty, *Wittig* proposed [170] two structures for the ylid and

$$\overset{+}{(CH_3)_3N}{-}\overset{-}{CH_2}LiBr$$

1 Trimethylammoniummethylid lithium bromide complex

$$\overset{+}{(CH_3)_3N}{-}CH_2{-}Li\ Br^-$$

1a Lithiomethyltrimethyl-ammonium bromide

presented evidence favoring structure *1* by noting that the characteristics of the ylid changed markedly if complexing with lithium bromide was prevented or made impossible. If the insoluble material obtained from the reaction of tetramethylammonium bromide with phenyllithium in ether is placed in 1,2-dimethoxyethane, a substance known to form a strong complex with the lithium ion, a rapid decomposition occurs to give trimethylamine and polymethylene in 74% yield. These products presumably arise from the decomposition of the free ylid after the stabilizing lithium bromide moiety is removed by complexation with solvent.

$$(CH_3)_3\overset{+}{N}-\overset{-}{C}H_2LiBr \quad + \quad CH_3OCH_2CH_2OCH_3$$

$$(CH_3)_3\overset{+}{N}-\overset{-}{C}H_2 \quad +$$

$$(CH_3)_3N \quad + \quad (CH_2)_n$$

It might be expected that the stability of the sodium bromide complex of the ylid would be much lower than the lithium bromide complex. Thus, when the ylid is prepared using a mixture of phenylsodium and phenyllithium, decomposition also occurs, but in this case the predominant product is dimethylethylamine [179].

Wittig reasoned that if formula *1a* more clearly represents the structure of the ylid, it would be expected that the halide ion would exert little influence on the reactivity of the ylid. It was found, however, that the product prepared by treatment of iodomethyltrimethylammonium iodide with phenyllithium in ether *did not react* with benzophenone, but the product from the reaction of bromomethyltrimethylammonium bromide with phenyllithium gave the same addition compound with benzophenone as the ylid prepared from tetramethylammonium bromide and phenyllithium [170]. Also, the reactivity of the ylid prepared from bromomethyltrimethylammonium bromide in ether was reduced when lithium iodide was added to the reaction mixture. This reduction in the reactivity of the ylid caused by the iodide ion was interpreted as good evidence that the halide ion is not free as indicated by structure *1a* but is closely associated with lithium as in structure *1* [170]. When chloromethyltrimethylammonium chloride is treated with phenyllithium, coupling occurs to give trimethylbenzylammonium chloride rather than the ylid [179]. The trimethylbenzylammonium salt subsequently undergoes a *Stevens* rearrangement to give dimethyl-α-methylbenzylamine in 22% yield (see Sect. VIA)

$$(CH_3)_3\overset{+}{N}-CH_2Cl\ Cl^- + C_6H_5Li \rightarrow (CH_3)_3\overset{+}{N}CH_2C_6H_5\ Cl^-$$

$$(CH_3)_2N-CHC_6H_5 \leftarrow (CH_3)_3\overset{+}{N}-\overset{-}{C}HC_6H_5$$
$$\quad\ |$$
$$\ \ CH_3$$

Subsequently, *Weygand, Daniel* and *Schroll* [158] reinvestigated the structure of the ylid and interpreted their results as indicating that structure *1a* is a more appropriate representation of the structure of the ylid. In their investigation, a variety of acylations of the ylid with carbon dioxide, benzonitrile, ethylbenzoate, acetyl chloride, and benzoyl chloride were carried out and the products obtained from these reactions were analogous to those obtained from similar acylations of lithium alkyls.

$$
[(CH_3)_3\overset{+}{N}\overset{-}{CH_2}Li]Br + R\overset{O}{\overset{\|}{C}}{-}X \;\rightarrow\; (CH_3)_3\overset{+}{N}{-}CH_2{-}\underset{\underset{O}{\|}}{C}{-}R \; X^-
$$

$$1a$$

In further efforts to show the similarities to lithium alkyls, the ylid was treated with diphenylmercury yielding phenyllithium by an exchange reaction [39].

$$
2\,(CH_3)_3\overset{+}{N}\overset{-}{CH_2}LiBr + (C_6H_5)_2Hg \;\rightarrow\; 2\,C_6H_5Li + ((CH_3)_3\overset{+}{N}CH_2)_2Hg + HgBr_2
$$

Such exchange reactions between mercurials and lithium alkyls and between mercurials and Grignard reagents had previously been demonstrated by *Gilman* and *Jones* [49] as well as by *Salinger* and *Dessy* [105]. In addition, *Daniel* and *Paetsch* [39] carried out a low temperature oxidation of the ylid with molecular oxygen in tetrahydrofuran, producing a soluble peroxide whose presence was detected both by iodometric titration with sodium thiosulfate and by the isolation of formaldehyde from the reaction mixture after hydrolysis of the formocholine intermediate *3* with acid.

$$
(CH_3)_3\overset{+}{N}\overset{-}{CH_2}LiBr + O_2 \xrightarrow{\;-110°\;} (CH_3)_3\overset{+}{N}\overset{-}{CH_2}OOLi
$$

$$H_2O \diagdown \qquad \diagup I^-$$

$$(CH_3)_3\overset{+}{N}CH_2OH\ OH^- \qquad\qquad I_2\ (Titrated)$$

$$3$$

$$HBr \diagdown$$

$$(CH_3)_3N + CH_2O + H_2O$$

The formation of peroxides by lithium alkyls and Grignard reagents, when treated with oxygen at low temperature, had been previously demonstrated by *Hock* and *Ernst* [68]. Finally, it was noted that lithium alkyls react with ethylene bromide to give off ethylene. Similarly, solutions of the ylid in tetrahydrofuran were found to react vigorously with ethylene bromide, giving off ethylene [39].

Although *Wittig* [170] noted that the reactivity of the ylid in ether was reduced when lithium iodide was added, *Daniel* and *Paetsch* [39] observed that the reactivity of the ylid was not reduced when lithium iodide was added to a THF solution of the ylid. In addition, they observed C—Li bond stretching vibrations in the infrared spectrum of a THF solution of the ylid [40]. The absorptions at 385, 425, 475, 500 and 580 cm^{-1}, which were attributed to the ylid, disappear on treatment with oxygen. The new band which appears at 450 cm^{-1} is attributable to an Li—O stretching frequency [40].

On the basis of their results, *Daniel* and *Paetsch* [39,40] concluded that *the correct formulation of the ylid* is $[(CH_3)_3N—CH_2Li]$ $^+Br^-$ since it behaves in a manner which is similar to known organolithium reagents. Indeed this formulation accurately describes the behavior of the ylid and perhaps is useful, but it also requires an ionic species in etherial solvents. However, it is known that lithium halides are highly associated in ethers [50] and it would be most surprising if this particular salt was not associated. Organolithium reagents are also highly associated in ethers [18] and appear to interact strongly with lithium halides [50,146].

It seems that the interpretation of an ionic ylid is akin to saying that the *Grignard reagent* might be formulated as $[RMg]$ $^+Br^-$ in etherial solvents. The Grignard reagent undergoes the same reactions as the lithium bromide complex of the ylid (exchange with mercurials, reaction with carbonyl compounds, etc.) even though the bromide is intimately associated with magnesium in THF [123,151]. Although the experimental observations on the lithium halide adducts of the ylid indicate that it reacts as an organolithium reagent, this may not preclude the presence of a bromide ion intimately associated with lithium in a manner which may be similar to the structure of the Grignard reagent. Using this formulation, the behavior of the lithium bromide complex can be compared with other *Lewis* acid complexes of the ylid. It would be interesting to examine the behavior of the tetrafluoroborate and tetraphenylborate salts of the ylid since interaction of this anion with the cation would be expected to be weak. However, this might not completely solve the dilemma since lithium perchlorate is highly associated in ethers [13].

B. Decomposition Reactions

If the free ylid, $(CH_3)_3\overset{+}{N}—\overset{-}{C}H_2$, is not stabilized by a *Lewis* acid, it decomposes. Even on standing in ether, the lithium bromide complex slowly decomposes to give trimethylamine and polymethylene [40]. The electronic structure of the free ylid has been discussed by *Cram* [35]. If the electrons occupy an sp^3 orbital in the free ylid, the charges would be slightly closer together than they would be if the electron pair occupied a pure p-orbital. If the electron pair was in an s-orbital the charges would be even closer together. Thus the electron pair in the free ylid may be in an orbital richer in s-character than sp^3. However, in stabilized ylids, the *Lewis* acid would be expected to be coordinated to an sp^3 hybridized carbon atom.

Investigations by *Wittig* [47] and by *Weygand* [158] indicate that ylids may give carbenes on decomposition. An initial report by *Franzen* and *Wittig* [47] claimed that a solution of the ylid prepared by a mixture of phenylsodium and phenyllithium in the presence of cyclohexene gave small yields of norcarane. Although *Wittig* and *Krauss* [179] were unable to repeat this work, (only 0.4% norcarane was isolated) they were able to isolate norcarane derivatives by using substituted quaternary ammonium salts. For instance, treatment of *n*-butoxymethyltrimethylammonium bromide *4* with phenyllithium in the presence of cyclohexene produced 7-(*n*-butoxy)norcarane.

$$C_4H_9OCH_2\overset{+}{N}(CH_3)_3 \ Br^- \ + \ \bigcirc \ \xrightarrow{C_6H_5Li} \ C_4H_9O\overset{}{\underset{H}{C}} \bigcirc \ + \ C_6H_6$$

$$4$$

$$+ \ (CH_3)_3N \ + \ LiBr$$

Similarly, 7-phenoxynorcarane was isolated from the reaction of phenoxymethyltrimethylammonium bromide with phenyllithium in the presence of cyclohexene. The isolation of these norcarane derivatives suggests the intermediate presence of a carbene or carbenoid precursor, presumably arising from decomposition of the initially formed ylid. It is worth noting that the reaction of alkylchloromethyl ethers with butyllithium also gives an alkoxycarbenoid derivative which inserts into olefins to give cyclopropanes [130]. *Franzen* and *Wittig* [47] also observed that treatment of tetramethylammonium chloride with phenyllithium: phenylsodium in THF in the presence of triphenylphosphine gave methylenetriphenylphosphorane. They concluded that this observation suggests the intermediate formation of a *carbene*.

$$[(CH_3)_3\overset{+}{N}CH_3]Cl^- + \phi Li : \phi Na + \phi_3P \xrightarrow{THF} (CH_3)_3N + \phi_3P{=}CH_2$$

Although no insertion reaction is observed in the presence of cyclo-hexene, the ylid decomposes to give dimethylethylamine (49%) rather than trimethylamine and polymethylene [179]. This amine results from a *Stevens* rearrangement of the free ylid.

$$(CH_3)_3 \overset{+}{N} - \overset{-}{CH_2} \rightarrow (CH_3)_2 NCH_2 CH_3$$

If the reaction between tetramethylammonium bromide and phenyl-lithium: phenylsodium is carried out in the absence of cyclohexene both dimethylethylamine and polymethylene are formed. Since no cyclohexene insertion product is observed, it was suggested [179] that polymethylene formation proceeds in the following way.

$$(CH_3)_3 \overset{+}{N} - \overset{-}{CH_2} + (CH_3)_3 \overset{+}{N} CH_3 \rightarrow (CH_3)_3 N + (CH_3)_3 \overset{+}{N} - CH_2 - CH_3$$

$$\xleftarrow{\text{etc.}} (CH_3)_3 \overset{+}{N} - CH_2 CH_2 CH_3 + (CH_3)_3 N \leftarrow \overset{}{\underset{(CH_3)_3 N^+ C^- H_2}{\Big\downarrow}}$$

The decomposition of the ylid in the presence of phenylsodium should be contrasted to the decomposition of the ylid with dimethoxyethane [169] where only polymethylene is observed (74%). It is worth noting that when dimethylsulfoniummethylide is decomposed in dimethylsulfoxide, ethylene is produced [34]; however, when this ylid is generated in ether by treatment of trimethylsulfonium bromide with phenyllithium, poly-methylene is formed in 70% yield [168]. It has also been shown that if lithium iodide is treated with diazomethane in ether, both polymethylene (12%) and ethylene (67%) are formed [180]. However, only a trace (0.4%) of norcarane is noted, when this decomposition is carried out in the presence of cyclohexene [180].

Wittig initially reported that the alkylation of the ylid with methyl iodide gave a low yield of trimethylethylammonium iodide [162]. Sub-sequently, *Weygand* and co-workers [158] found that the reaction of completely ^{14}C-labeled ylid with unlabeled methylbromide resulted in the formation of both tetramethylammonium bromide and trimethyl-ethylammonium bromide in a 5 to 1 ratio. These two products contained only about 80% of the original [14]C. The remainder of the ^{14}C was found in the polymethylene which had formed during the reaction. To explain these results, it was suggested that most of the ylid decomposes to give trimethylamine and polymethylene before it can be alkylated by methyl bromide.

$$(CH_3)_3\overset{+}{N}CH_2\overset{-}{L}iBr \xrightarrow[16\%]{CH_3Br} CH_3CH_2\overset{+}{N}(CH_3)_3Br^-$$

84%

$$-(CH_2)_{\overline{n}} \leftarrow [:CH_2] + (CH_3)_3N \xrightarrow{CH_3Br} (CH_3)_4\overset{+}{N}Br^-$$

An alternative decomposition route leading to polymethylene can also be envisioned where the ylid acts as a proton abstracting agent to induce an α-elimination.

$$(CH_3)_3\overset{+}{N}-\overset{-}{C}H_2LiBr + CH_3Br \rightarrow (CH_3)_4\overset{+}{N} + [CH_2] + Br^-Li^+$$

$$n\,[CH_2] \rightarrow -(CH_2)_n-$$

This would account for the observation that the polymethylene contains less carbon-14 than expected if only the preceding mechanism was operating. In reactions of the ylid with other alkyl bromides, proton abstraction occurs in preference to substitution. Reaction with cyclohexylbromide gives a 92% yield of cyclohexene by a β-elimination. Reaction with benzyl bromide gives a 58% yield of stilbene, probably by α-elimination, with only a small amount of 2-phenylethyltrimethylammonium bromide by substitution [158].

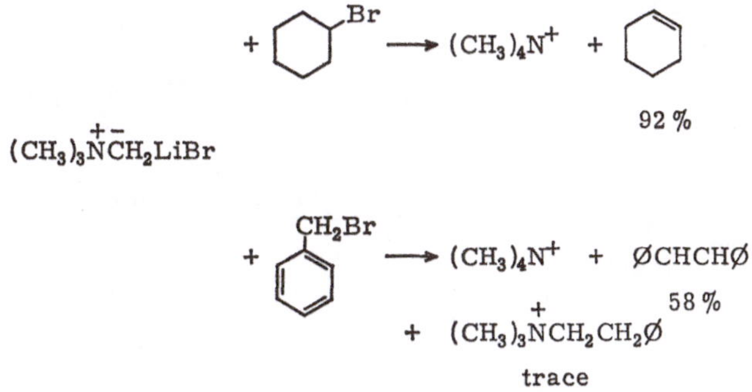

$$+ \underset{}{\bigcirc}^{Br} \longrightarrow (CH_3)_4N^+ + \bigcirc$$

92 %

$$(CH_3)_3\overset{+}{N}\overset{-}{C}H_2LiBr$$

$$+ \underset{}{\bigcirc}^{CH_2Br} \longrightarrow (CH_3)_4N^+ + \emptyset CHCH\emptyset$$

58 %

$$+ (CH_3)_3\overset{+}{N}CH_2CH_2\emptyset$$

trace

In extensive studies of α-elimination reactions, *Wittig* [176,180], *Simmons* and *Smith* [136], *Köbrich* [85], and *Closs* [51], have observed carbenoid behavior in reactions of various α-haloorganometallics.

$$Cl-HC\begin{matrix} Cl \\ \\ Li \end{matrix} \qquad H_2C\begin{matrix} I \\ \\ MgI \end{matrix} \qquad H_2C\begin{matrix} I \\ \\ ZnI \end{matrix} \qquad H_2C\begin{matrix} N_2^+ \\ \\ Li \end{matrix}$$

Although these carbenoids are usually discussed in relation to insertion reactions, some of them undergo polymerization and other reactions which are similar to those of the ylid. Thus, in ylid chemistry the $(CH_3)_3N^+$ group *may be considered as a pseudo-halogen*. Although it has not been shown that the ylid reacts by an insertion reaction, it is possible that the conditions under which insertion can occur have not been realized. If the ylid is considered as a carbenoid, its polymerization reactions may proceed *via* a lithium halide complex. Alternatively, the complex may rearrange to the bromomethyllithium which may be the reactive intermediate.

$$H_2\bar{C}\begin{matrix} N^+ \\ + \\ LiBr \end{matrix} \rightarrow \left[-\overset{|}{\underset{|}{N}}: \quad CH_2-LiBr \right] \rightarrow Br-CH_2-Li$$

Compare this rearrangement with a possible rearrangement path of organoborane complexes. (See Sect. II. D. 1.)

$$-\overset{|}{\underset{|}{N^+}}-CH_2-\overset{R}{\underset{R}{\overset{|}{\underset{|}{B}}}}-\bar{R} \rightarrow \left[-\overset{|}{\underset{|}{N}}: \quad CH_2-\overset{R}{\underset{R}{\overset{|}{\underset{|}{B}}}}-R \right] \rightarrow \overset{R}{\underset{|}{CH_2}}-BR_2$$

The closest analogy is the mechanism proposed for the lithium halide catalyzed decomposition of diazomethane in ether to give ethylene and polymethylene [180].

$$CH_2{=}N{=}N + Li^+ \rightarrow H_2C\begin{matrix} \overset{+}{N}{\equiv}N \\ \\ Li \end{matrix} \xrightarrow{+X^-} H_2C\begin{matrix} X \\ \\ Li \end{matrix} + N_2$$

$$polymethylene \xleftarrow[\text{or } XCH_2Li]{CH_2N_2} H_2C\begin{matrix} \overset{+}{CH_2{-}N_2} \\ \\ Li \end{matrix} \xleftarrow{CH_2N_2 \text{ or } XCH_2Li}$$

$$\xrightarrow[-N_2]{-Li^+} CH_2{=}CH_2$$

This path can also be visualized in the fluorenylide system [161)]

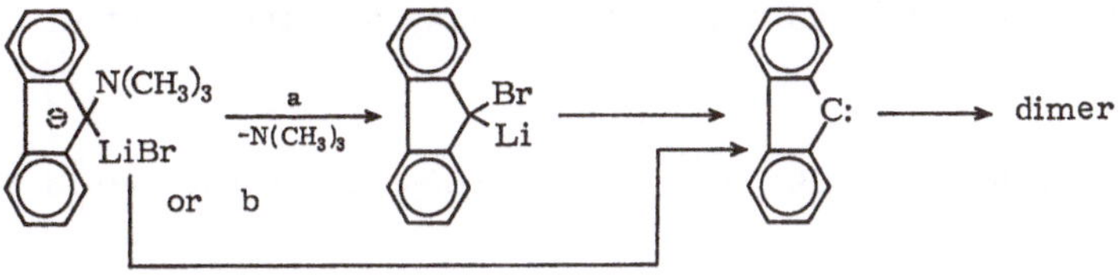

If this rearrangement occurs, polymerization and dimerization reactions of the ylid or the bromoalkyllithium reagent can be considered in the following way.

1   or $BrCH_2Li + \equiv \overset{+}{N} - \overset{-}{C}H_2LiBr \rightarrow \equiv \overset{+}{N} - CH_2 - CH_2 - \overset{-}{LiBr} +$
$$(CH_3)_3N + LiBr$$

1 or $BrCH_2Li + \equiv \overset{+}{N} - CH_2 - \overset{-}{C}H_2 - LiBr \rightarrow \equiv \overset{+}{N} - CH_2 - CH_2 - \overset{-}{C}H_2 - LiBr +$
$$(CH_3)_3N + LiBr$$

$$\equiv \overset{+}{N} - CH_2\overset{-}{C}H_2LiBr \rightarrow \equiv N : + CH_2 = CH_2 + LiBr$$

1   or $BrCH_2Li \rightarrow [CH_2] \rightarrow$ insertion products?

Daniel and *Paetsch* [40)] carried out a series of labeling experiments designed to determine whether the methylene carbon atom of the ylid retains its integrity during decomposition. The ylid was labeled in several positions with ^{14}C and ^3H, but the specific ylid prepared from bromomethyltrimethylammonium bromide and methyllithium with the methylene group labeled with ^{14}C will be discussed here.

$$[(CH_3)_3\overset{+}{N}\overset{*}{C}H_2Br]\ Br^- \xrightarrow{CH_3Li} (CH_3)_3\overset{+}{N} - \overset{-}{\overset{*}{C}}H_2LiBr \xrightarrow{35°} (CH_3)_3N + \overset{*}{(}CH_2)_n$$

If the ylid prepared in this manner is allowed to decompose in ether at 35 °C, no ^{14}C is observed in the trimethylamine. Therefore, the carbon atoms in the ylid do not equilibrate under these conditions. However, if the ylid is prepared by treatment of the labeled bromomethyltrimethyl-

307

ammonium bromide with a mixture of phenyllithium and phenylsodium, the ylid decomposes with the label 60% equilibrated throughout all the methyl groups. Under these conditions the ylid decomposes *via* a *Stevens* rearrangement to give dimethylethylamine and the extent of carbon

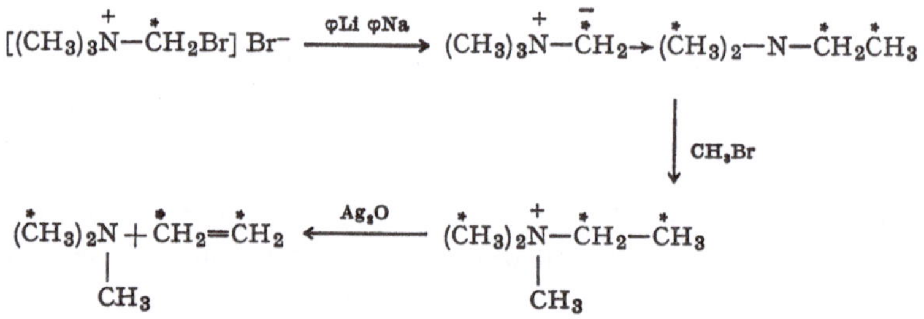

$$[(CH_3)_3\overset{+}{N}-\overset{\bullet}{C}H_2Br]\ Br^- \xrightarrow{\ \phi Li\ \phi Na\ } (CH_3)_3\overset{+}{N}-\overset{\bullet}{C}H_2 \rightarrow (CH_3)_2-N-\overset{\bullet}{C}H_2\overset{\bullet}{C}H_3$$

$$\downarrow CH_3Br$$

$$\underset{\underset{CH_3}{|}}{(\overset{\bullet}{C}H_3)_2\overset{\bullet}{N}} + \overset{\bullet}{C}H_2{=}\overset{\bullet}{C}H_2 \xleftarrow{\ Ag_2O\ } \underset{\underset{CH_3}{|}}{(\overset{\bullet}{C}H_3)_2\overset{+}{N}-\overset{\bullet}{C}H_2-\overset{\bullet}{C}H_3}$$

scrambling was ascertained by determining the amount of ^{14}C in the trimethylamine formed in the analytical scheme shown above. Since scrambling occurs, both an internal S_Ni mechanism [81] and a simple ion-pair mechanism [76] were rejected and it was suggested [40] that the immonium ion pair below is the intermediate which leads to the *Stevens rearrangement product.*

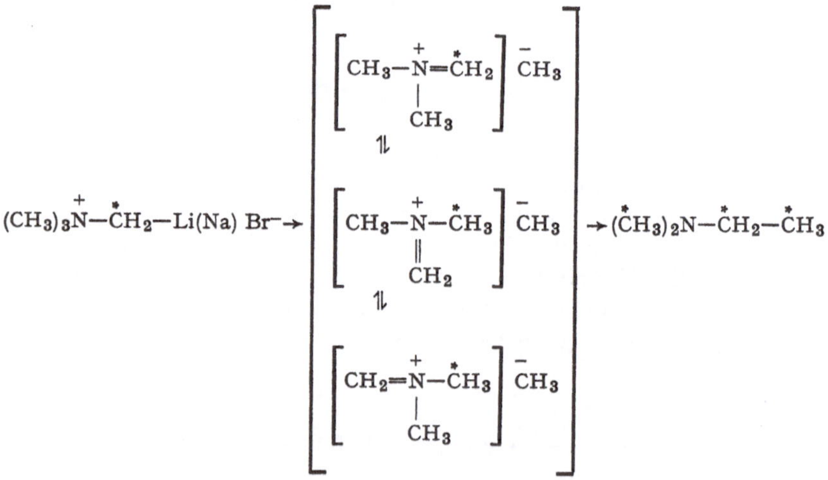

$$(CH_3)_3\overset{+}{N}-\overset{\bullet}{C}H_2-Li(Na)\ Br^- \rightarrow \left[\begin{array}{c} \left[\underset{\underset{CH_3}{|}}{CH_3-\overset{+}{N}{=}\overset{\bullet}{C}H_2}\right] \overset{-}{C}H_3 \\ \Updownarrow \\ \left[\underset{\underset{CH_2}{\|}}{CH_3-\overset{+}{N}-\overset{\bullet}{C}H_3}\right] \overset{-}{C}H_3 \\ \Updownarrow \\ \left[\underset{\underset{CH_3}{|}}{CH_2{=}\overset{+}{N}-\overset{\bullet}{C}H_3}\right] \overset{-}{C}H_3 \end{array} \right] \rightarrow (\overset{\bullet}{C}H_3)_2N-\overset{\bullet}{C}H_2-\overset{\bullet}{C}H_3$$

In this scheme an intramolecular proton exchange in the immonium ion is apparently required to equilibrate the carbon atoms. However, an alternative mechanism involving an intramolecular proton exchange

[45,109] in the ylid prior to rearrangement by any of the possible mechanisms would also lead to scrambling in the dimethylethylamine and account for the experimental observations. Further discussion of intramolecular

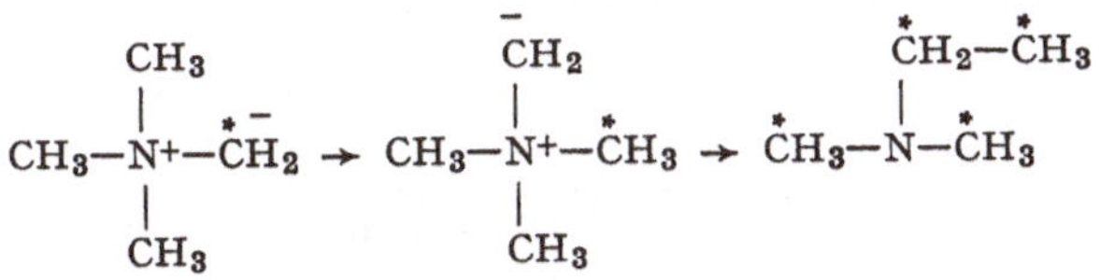

$$
\begin{array}{ccc}
CH_3 & \overset{-}{CH_2} & \overset{\bullet\,\bullet}{CH_2}-CH_3 \\
| & | & | \\
CH_3-\overset{+}{N}-\overset{\bullet\;-}{CH_2} \;\rightarrow\; CH_3-\overset{+}{N}-\overset{\bullet}{CH_3} \;\rightarrow\; \overset{\bullet}{CH_3}-N-\overset{\bullet}{CH_3} \\
| & | & \\
CH_3 & CH_3 &
\end{array}
$$

exchange reactions in the ylid is presented in Sect. III A.

C. Condensation and Displacement Reactions

The ylid can be considered as an organolithium reagent in many of its reactions. The reactions with benzophenone and other compounds containing an electrophilic carbon illustrate this behavior [158].

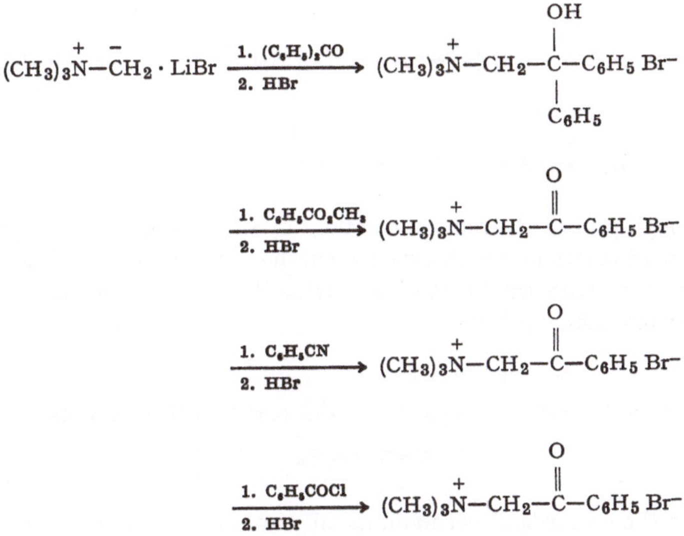

$$
(CH_3)_3\overset{+}{N}-\overset{-}{CH_2}\cdot LiBr \xrightarrow[\text{2. HBr}]{\text{1. } (C_6H_5)_2CO} (CH_3)_3\overset{+}{N}-CH_2-\underset{\underset{C_6H_5}{|}}{\overset{\overset{OH}{|}}{C}}-C_6H_5 \; Br^-
$$

$$
\xrightarrow[\text{2. HBr}]{\text{1. } C_6H_5CO_2CH_3} (CH_3)_3\overset{+}{N}-CH_2-\overset{\overset{O}{\|}}{C}-C_6H_5 \; Br^-
$$

$$
\xrightarrow[\text{2. HBr}]{\text{1. } C_6H_5CN} (CH_3)_3\overset{+}{N}-CH_2-\overset{\overset{O}{\|}}{C}-C_6H_5 \; Br^-
$$

$$
\xrightarrow[\text{2. HBr}]{\text{1. } C_6H_5COCl} (CH_3)_3\overset{+}{N}-CH_2-\overset{\overset{O}{\|}}{C}-C_6H_5 \; Br^-
$$

Although these reactions occur initially, secondary reactions are often observed. During the reaction of the ylid with benzoyl chloride, the following series of reactions takes place [135].

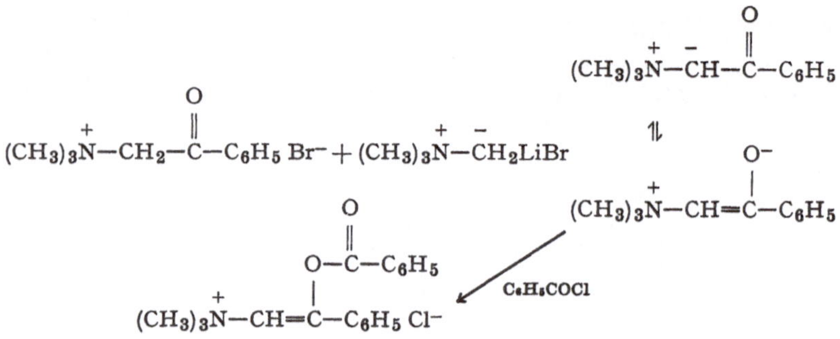

$$(CH_3)_3\overset{+}{N}-CH_2-\overset{O}{\overset{\|}{C}}-C_6H_5\ Br^- + (CH_3)_3\overset{+}{N}-\overset{-}{C}H_2LiBr$$

$$(CH_3)_3\overset{+}{N}-\overset{-}{C}H-\overset{O}{\overset{\|}{C}}-C_6H_5$$

$$\updownarrow$$

$$(CH_3)_3\overset{+}{N}-CH=\overset{O^-}{\overset{|}{C}}-C_6H_5$$

$$\overset{O}{\overset{\|}{O-C}}-C_6H_5$$

$$(CH_3)_3\overset{+}{N}-CH=\overset{|}{C}-C_6H_5\ Cl^-$$

$$\xleftarrow{C_6H_5COCl}$$

No acylation at the methylene carbon is observed.

A similar sequence of reactions occurs with carbon dioxide. The first product, betaine, is the primary product if the ylid is added to CO_2, but,

$$(CH_3)_3\overset{+}{N}-\overset{-}{C}H_2LiBr\ +\ CO_2\ \rightarrow\ (CH_3)_3\overset{+}{N}-CH_2CO_2^-$$

if the carbon dioxide is added to a solution of ylid in THF, the betaine reacts further to give trimethylammoniomalonate [155].

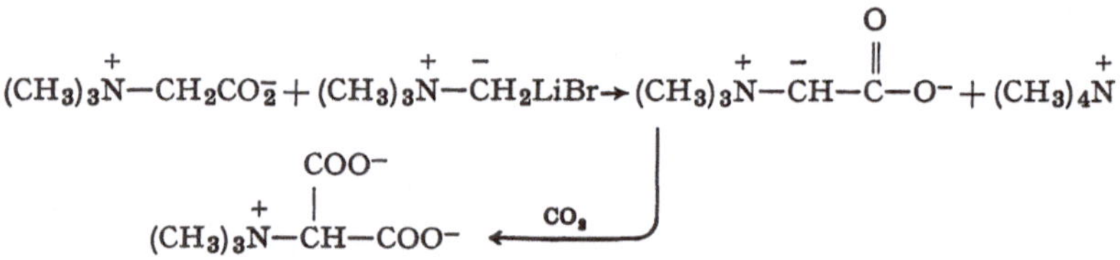

$$(CH_3)_3\overset{+}{N}-CH_2CO_2^- + (CH_3)_3\overset{+}{N}-\overset{-}{C}H_2LiBr \rightarrow (CH_3)_3\overset{+}{N}-\overset{-}{C}H-\overset{O}{\overset{\|}{C}}-O^- + (CH_3)_4\overset{+}{N}$$

$$\overset{COO^-}{(CH_3)_3\overset{+}{N}-\overset{|}{C}H-COO^-} \xleftarrow{CO_2}$$

Although proton abstraction reactions often occur with alkyl halides, displacement reactions are observed when the ylid is treated with organometallics, e.g., trimethylbromosilane, trimethylbromogermane, and trimethylbromostannane [110].

$$(CH_3)_3\overset{+}{N}-\overset{-}{C}H_2LiBr + (CH_3)_3MBr \xrightarrow{THF} (CH_3)_3\overset{+}{N}CH_2-M(CH_3)_3\ Br^-$$

$$M=Si,\ Ge,\ Sn$$

When the silicon and germanium salts are heated, N-demethylation occurs and the products are methyl bromide and dimethylaminomethyltrimethylsilane.

$$(CH_3)_3\overset{+}{N}-CH_2Si(CH_3)_3\ Br^- \xrightarrow{270\ °C} CH_3Br + (CH_3)_2NCH_2Si(CH_3)_3$$

However, when the tin salt is heated, trimethylbromostannane, trimethylamine and polymethylene are formed. Because the tin salt decomposes in this way it may not be ionic, as written above, and could be considered as a pentacovalent compound of tin [110]. If this alternative structure is correct,

$$(CH_3)_3\overset{+}{N}-CH_2-\underset{\underset{CH_3}{|}}{\overset{\overset{\displaystyle H_3C}{\diagdown}\diagup\overset{\displaystyle CH_3}{}}{Sn}}-Br \xrightarrow{270\ °C} (CH_3)_3N + (CH_2)_n + (CH_3)_3SnBr$$

its properties may be similar to the lithium bromide complex, with $(CH_3)_3Sn^+$ replacing Li^+. Perhaps carbenoid behavior can be observed with this ylid complex, but more work needs to be done. Similar derivatives of methylenetriphenylphosphorane have also been prepared by *Seyferth* and *Grim* [132] but their decomposition reactions have not been reported.

D. Adduct Formation and Transmetalation Reactions

Because of its similarity to trimethylamine oxide, the ylid may be considered as a *Lewis* base, and, as such, it should coordinate with various *Lewis* acids. Complexes with lithium bromide and trimethylbromostannane have already been mentioned and it would be expected that coordination with boranes and alanes would also give stable products.

1. Boron Containing Adducts

Since the boron trifluoride adduct is the most stable adduct of the boron complexes of methylenetriphenylphosphorane [121], it might be expected that this *Lewis* acid would also form a stable nitrogen ylid adduct. Treatment of the ylid with boron trifluoride gave a white crystalline solid whose properties were consistent with the proposed adduct [110]. The adduct decomposes to trimethylamine trifluoroborane

$$(CH_3)_3\overset{+}{N}-\overset{-}{CH_2}LiBr + BF_3 \rightarrow (CH_3)_3\overset{+}{N}-CH_2-\overset{-}{B}F_3$$

$$(CH_3)_3NBF_3 + CH_2=CH_2 \longleftarrow \Bigg] 275°$$

and ethylene on heating. In examining the nmr and conductivity of this adduct in nitromethane, it appears that the salt may be partly ionized in this solvent. Notice that the cation produced on ionization is isoelectronic with betaine,

$$(CH_3)_3\overset{+}{N}-CH_2-\overset{-}{B}F_3 \rightleftharpoons (CH_3)_3\overset{+}{N}-CH_2-BF_2\ F^-$$

$(CH_3)_3\overset{+}{N}-CH_2CO_2^-$, the carbon dioxide adduct of the ylid, and may be expected to be stable. In an attempted preparation [110] of the boron trichloride adduct, an extremely unstable compound was isolated which had an nmr spectrum consistent with the expected adduct, but could not be purified. The BCl_3 adduct of methylenetriphenylphosphorane is also unstable and cannot be purified [121].

Since *Seyferth* and *Grim* [133] observed that trialkylborane adducts of methylenetriphenylphosphorane were unstable, *Musker* and *Stevens* [106] attempted to demonstrate the existence of the adduct, not by the isolation of the adduct itself, but rather by the isolation of reaction products whose formation might be rationalized by a mechanism involving an ylid-borane adduct [106].

When a solution of the ylid in THF was treated with a trialkylborane, trimethylamine was evolved from the reaction mixture. In view of the facile elimination of trimethylamine and the knowledge that trimethylamine oxide can be used to oxidize organoboranes [87] — a process which is thought to involve the migration of a boron substituent to oxygen — it was suggested that a similar migration had occurred with the isoelectronic nitrogen ylid [106]. After the evolution of trimethylamine ceased, the mixture was oxidized with

$$(CH_3)_3\overset{+}{N}-\overset{-}{C}H_2LiBr + BR_3 \rightarrow (CH_3)_3\overset{+}{N}-CH_2-\overset{-}{B}R_3$$

$$\begin{array}{c} CH_2R \\ | \\ (CH_3)_3N + BR_2 \end{array} \Bigg\rfloor$$

$$RCH_2BR_2 \xrightarrow[OH^-]{H_2O_2} RCH_2OH + 2\ ROH + B(OH)_3$$

alkaline hydrogen peroxide and the alcohols which were formed were analyzed. With trihexylborane, a 63% yield of heptanol (based on the migration of one R group) was obtained. When *bis*(3-methyl-2-butyl)-

hexylborane was used as the *Lewis* acid, a 60% yield of 2,3-dimethyl-1-butanol and a 6% yield of heptanol was obtained. These results show that secondary alkyl groups migrate more readily than primary alkyl groups. The phenyl group also migrated from boron to carbon (46% yield of benzyl alcohol from triphenylborane), but no hydrogen migration was ever observed (0% methanol from either *bis*(3-methyl-2-butyl) borane or diborane) [106,110]. Although a simple 1—2 migration satisfactorily accounts for the observed results, a carbene mechanism is a possible alternative [106]. *Tufariello* and *Lee* [149] have shown that similar results are obtained when the sulfur ylid, dimethylsulfoxoniummethylide, is treated with organoboranes. Although organoborane adducts of methylenetriphenylphosphorane are more stable than those of the nitrogen ylid, *Köster* and *Rickborn* [88] reported that the migration of a phenyl group in triphenylphosphinemethylene triphenylborane can be realized if the adduct is heated in decalin. One organoborane adduct of the nitrogen ylid has recently been obtained in pure form [15]. The crystalline triphenylborane adduct of *1* can be isolated if the lithium bromide adduct is treated with triphenylborane in THF at room temperature. However, if the adduct is heated in 1,2-dimethoxyethane (DME) at

$$(CH_3)_3\overset{+}{N}-\overset{-}{CH_2}\cdot LiBr + (C_6H_5)_3B \xrightarrow{\text{THF}} (CH_3)_3\overset{+}{N}-CH_2-\overset{-}{B}(C_6H_5)_3$$

83 °C, phenyl migration can be caused to occur. One of the interesting

$$(CH_3)_3\overset{+}{N}-CH_2-\overset{-}{B}(C_6H_5)_3 \xrightarrow[\text{DME}]{83\ °C} (CH_3)_3N + C_6H_5CH_2-B(C_6H_5)_2$$

reactions of this adduct, as reported by *Bickelhaupt* and *Barnick* [15], is the reaction with acetic acid containing a small amount of hydrochloric acid. The boronic acid which forms is extremely acidic (pK = 5.6) when

$$(CH_3)_3\overset{+}{N}-CH_2\overset{-}{B}(C_6H_5)_3 \xrightarrow[\text{HCl}]{CH_3COOH} [(CH_3)_3\overset{+}{N}-CH_2-B(OH)_2]\ Cl^- + 3\ C_6H_6$$

compared with common boronic acids (pK = 10 ~ 11) and ionizes to give a species which is isoelectronic with betaine.

$$(CH_3)_3\overset{+}{N}-CH_2-\overset{-}{B}\overset{\displaystyle O}{\underset{\displaystyle OH}{<}}$$

A related compound, piperidinomethaneboronic acid, has been prepared and then was converted to its catecholate ester and methylated with methyliodide to give the betaine shown below [104].

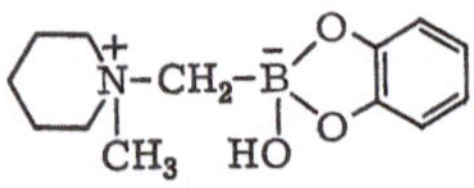

The dimethylaminomethaneboronic acid was prepared as its catecholate ester but was not methylated [104].

Since hydrogen migration was not observed with borane adducts, an attempt was made to prepare the BH_3 adduct of the ylid [110]. Although no trimethylamine was evolved from the reaction mixture, no adduct could be isolated.

2. Aluminum Containing Adducts

Like trialkylboranes, trialkylaluminum compounds are good *Lewis* acids which combine readily with such donors as amines, phosphines, ethers, and thioethers to give tetrahedral, four coordinated complexes that are more stable to dissociation than the corresponding trialkylborane complexes. Thus, the simplest trialkylaluminum compound, trimethylaluminum, forms a stable complex with the ylid [110]. Since the trialkylborane adducts of trimethylammoniummethylide, methylenetriphenyl-

$$(CH_3)_3\overset{+}{N}-\overset{-}{CH_2} \cdot LiBr + Al(CH_3)_3 \rightarrow (CH_3)_3\overset{+}{N}CH_2\overset{-}{Al}(CH_3)_3$$

phosphorane and dimethylsulfoxoniummethylide apparently undergo rearrangement by alkyl group migration on heating, the migration aptitude of the alkyl groups on the trialkylaluminum adduct of trimethylammoniummethylide was studied. When the adduct is heated to 160 °C for 12 hours and hydrolyzed with concentrated HCl, a mixture of hydrocarbon gases consisting of 78% methane, 19% ethane, and 2.9% propane is obtained, indicating that methyl group migration had occurred. From this data it can be calculated that about 60% rearrangement took place. This is approximately the same yield of rearrangement products that is observed in the reactions of organoboranes.

The formation of propane may be accounted for by the establishment of an equilibrium between the ylid and trimethylaluminum in the melted solid. This would allow interaction of the initially formed dimethylethylaluminium with the free ylid, effecting a second alkyl group migration to give dimethylpropylaluminum, which on hydrolysis gives propane [110].

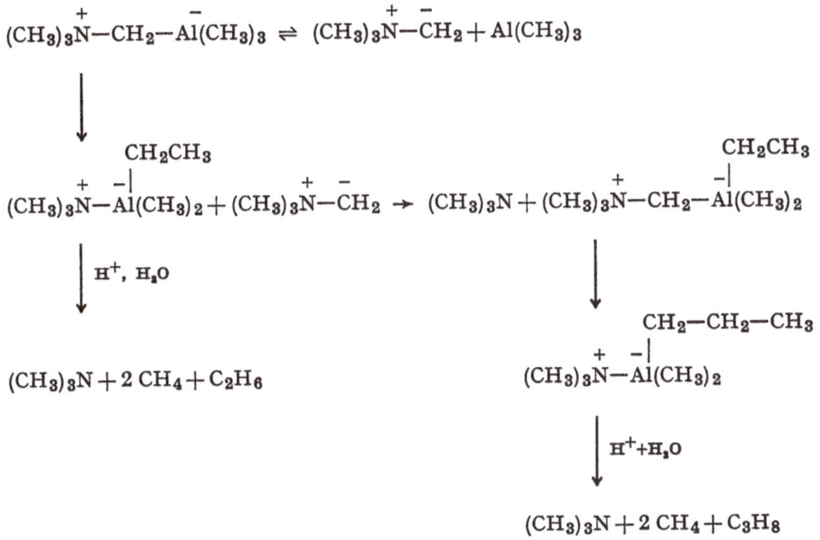

$$(CH_3)_3\overset{+}{N}-CH_2-\overset{-}{Al}(CH_3)_3 \rightleftharpoons (CH_3)_3\overset{+}{N}-\overset{-}{CH_2} + Al(CH_3)_3$$

$$\downarrow$$

$$\underset{(CH_3)_3\overset{+}{N}-\overset{-}{\underset{|}{Al}}(CH_3)_2}{\overset{CH_2CH_3}{}} + (CH_3)_3\overset{+}{N}-\overset{-}{CH_2} \rightarrow (CH_3)_3N + \underset{(CH_3)_3\overset{+}{N}-CH_2-\overset{-}{\underset{|}{Al}}(CH_3)_2}{\overset{CH_2CH_3}{}}$$

$$\downarrow H^+, H_2O \qquad\qquad\qquad\qquad \downarrow$$

$$(CH_3)_3N + 2\,CH_4 + C_2H_6 \qquad\qquad \underset{(CH_3)_3\overset{+}{N}-\overset{-}{\underset{|}{Al}}(CH_3)_2}{\overset{CH_2-CH_2-CH_3}{}}$$

$$\downarrow H^+ + H_2O$$

$$(CH_3)_3N + 2\,CH_4 + C_3H_8$$

Here again a carbenoid mechanism may be invoked. An attempt to form a trihexylaluminum adduct of the ylid was unsuccessful presumably due to the increased steric requirement of the hexyl group. Similar polymerization reactions have been observed by *Tufariello* and *Lee* [149] using organoborane adducts of dimethyloxosulfonium methylide.

It is worth noting that *Schmidbaur* and *Tronich* [127] recently prepared a trimethylaluminum adduct of methylenetriphenylphosphorane but its decomposition

$$\phi_3P=CH_2 + (CH_3)_3Al \rightarrow \phi_3\overset{+}{P}-CH_2-\overset{-}{Al}(CH_3)_3$$

was not reported.

3. Mercury and Zinc Containing Adducts

Daniel and *Paetsch* carried out a reaction of the lithium bromide complex of trimethylammoniummethylide with diphenylmercury to clarify the structure of the ylid in solution. Previously *Gilman* and *Jones* [49] as well as *Salinger* and *Dessy* [122)] have demonstrated, that in a solution containing both lithium and mercury alkyls, an equilibrium is established in which the organic carbanions of the two alkyls are exchanged. Thus, *Daniel* and *Paetsch* [39)], demonstrated that phenyllithium is formed (65% yield) in a reaction between the ylid and diphenylmercury and suggested that the ylid in solution reacts simply as a lithium alkyl; no attempt was made to isolate the mercurial.

When methylenetriphenylphosphorane is treated with mercuric bromide, a compound containing a carbon-mercury bond is obtained [132)].

$$HgBr_2 + 2\ (C_6H_5)_3\overset{+}{P}CH_2 \rightarrow Hg[CH_2\overset{+}{P}(C_6H_5)_3]_2Br_2^{\overline{}}$$

$$\downarrow \text{HgBr}_2 \text{ in CH}_3\text{OH}$$

$$[(C_6H_5)_3\overset{+}{P}-CH_2-Hg-CH_2-\overset{+}{P}(C_6H_5)_3]\ [HgBr_3^{\overline{}}]_2$$

Wittig and *Schwarzenbach* [176)] prepared the *bis*-ylid adduct of mercury by two different paths: 1) by reaction of *bis*-chloromethyl mercury with trimethylamine and 2) by reaction of the ylid with mercuric chloride.

In an attempt to isolate a similar compound by reaction of the ylid with diphenylmercury an amorphous solid was obtained whose infrared spectrum and pyrolysis products were consistent with the expected *bis*-ylid but which did not give a satisfactory analysis. The zinc adduct was also prepared [176)] from *bis*-chloromethylzinc and trimethylamine and from the ylid and zinc chloride, but zinc derivatives were not isolated from the reaction of the ylid with dipentyl zinc [110)].

E. Dimetalation Reactions

A complicating factor arises in the preparation of the ylid from tetramethylammonium bromide and phenyllithium if the lithium reagent is used in a large excess, or if a small excess is allowed to remain in contact with the ammonium salt for extended periods of time. *Wittig* and *Reiber* [165)] found that treatment of tetramethylammonium bromide with two equivalents of phenyllithium in ether produced a product mixture which reacted with benzophenone in tetrahydrofuran to yield a monoaddition compound as well as a *bis*-addition compound after hydrolysis with hydrobromic acid.

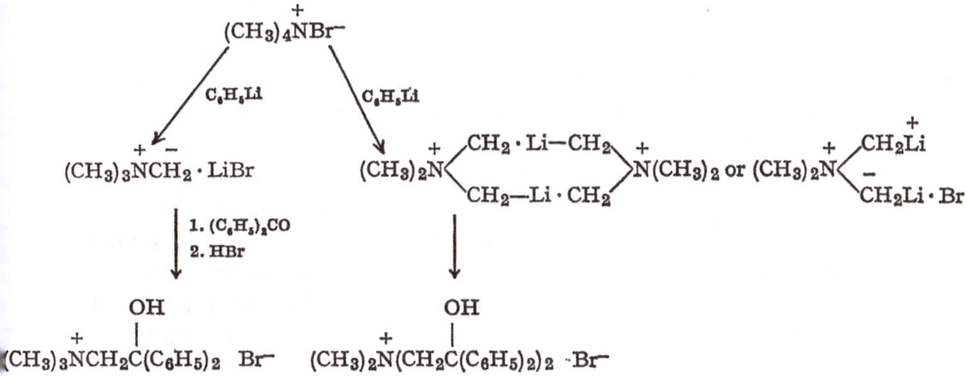

$$(CH_3)_4\overset{+}{N}Br^-$$

$$C_6H_5Li \qquad C_6H_5Li$$

$$(CH_3)_3\overset{+}{N}CH_2^- \cdot LiBr$$

$$(CH_3)_2\overset{+}{N}\underset{CH_2-Li\cdot CH_2}{\overset{CH_2\cdot Li-CH_2}{<}}\overset{+}{N}(CH_3)_2 \text{ or } (CH_3)_2\overset{+}{N}\underset{CH_2Li\cdot Br}{\overset{CH_2Li}{<}}^-$$

1. (C₆H₅)₂CO
2. HBr

$$\underset{(CH_3)_3\overset{+}{N}CH_2\overset{|}{\underset{}{C}}(C_6H_5)_2}{\overset{OH}{}} Br^-$$

$$\underset{(CH_3)_2\overset{+}{N}(CH_2\overset{|}{\underset{}{C}}(C_6H_5)_2)_2}{\overset{OH}{}} Br^-$$

Further investigation showed that the species thought to be a *bis*-ylid is quite soluble in ether whereas the mono-ylid is ether-insoluble. Increasing the time of contact of equivalent amounts of phenyllithium and the ammonium salt before addition of benzophenone led to a decrease in the amount of monoaddition compound formed. This appears to indicate a gradual conversion of the mono-ylid into the *bis*-ylid. More recently, *Wittig* and *Tochtermann* [175] demonstrated that *bis*-ylid formation is a general process by isolating the *bis*-adduct of benzophenone with the ylid prepared from N-methyl-N-bromomethylpyrrolidinium bromide with excess butyllithium at room temperature. If the ylid is prepared and reacted at −70 °C, only the mono-adduct of benzophenone is formed. *Bis*-ylids have also been proposed in reactions of benzyl substituted quaternary ammonium salts. (See Sect. VI A)

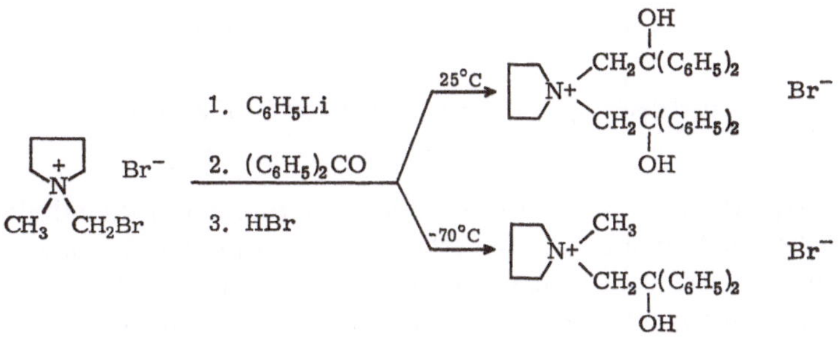

III. Reactions of the Tetramethylammonium Ion with Other Strong Bases

A. Hydroxide and Alkoxides

In 1881, *Hofmann* [69] reported that the thermal decomposition of tetra-methylammonium hydroxide gives trimethylamine and methanol. In 1964, a study by *Musker* [105] showed that the products resulting from the "dry" decomposition of tetramethylammonium hydroxide at 135—140 °C were trimethylamine and dimethylether. Only a trace of methanol was observed. Trace amounts of dimethyl ether was reported as a product of the pyrolysis of cyclohexylmethyl-β-d-trimethylammonium hydroxide and its formation was explained by a three step mechanism involving a sequence of S_N2 reactions [31]. An analogous mechanism for the decomposition of tetramethylammonium hydroxide, which would account for the observed products, was proposed by *Tanaka, Dunning*, and *Carter* [142].

$$(CH_3)_4\overset{+}{N}\,OH^- \rightarrow (CH_3)_3N + CH_3OH$$

$$CH_3OH + OH^- \rightarrow CH_3O^- + H_2O$$

$$CH_3O^- + (CH_3)_4\overset{+}{N} \rightarrow CH_3OCH_3 + (CH_3)_3N$$

Although this mechanism adequately accounts for the products, it is somewhat surprising that more methanol is not observed. Since strong bases have been shown to cause proton abstraction from the tetramethyl-ammonium ion, and other quaternary ammonium ions undergo hydrogen exchange with hydroxide in concentrated solution, an alternative ylid mechanism can be considered. Decomposition of the resulting ylid in the presence of a minimum amount of water could give the observed products.

$$(CH_3)_4\overset{+}{N}\,\overset{-}{OH} \rightarrow (CH_3)_3\overset{+}{N}-\overset{-}{CH_2} + H_2O$$

$$(CH_3)_3\overset{+}{N}-\overset{-}{CH_2} + \tfrac{1}{2}H_2O \rightarrow (CH_3)_3N + CH_3OCH_3$$

Perhaps the best rationalization for the formation of dimethylether rather than methanol is that the ether is the first product formed which could be evolved from a highly basic polar medium. Both water (hydroxide) or methanol (methoxide) would be expected to be retained in the solid (or semi-solid) mass. Methanol is the exclusive product, if the decomposition is carried out in the presence of excess water in a sealed tube at 270 °C [142].

$$(CH_3)_3\overset{+}{N}-CH_3 + OH^- \xrightarrow[\text{H}_2\text{O}]{270\ ^\circ\text{C}} (CH_3)_3N + CH_3OH$$

In a study of the hydrogen exchange between the tetramethylammonium, tetramethylphosphonium and trimethylsulfonium ions and D_2O at 100 °C in the presence of dilute base (0.3 M), *Doering* and *Hoffman* [43] observed that exchange in the phosphonium and sulfonium salts readily occurs whereas negligible exchange (1.5 atm % in 358 hrs) occurs in the ammonium salt. They attributed the facile exchange in the phosphonium and sulfonium salts to stabilization of the ylid (anion) by the d-orbitals of the central element thus making the protons more acidic. Since the nitrogen atom does not have d-orbitals available, the ylid is not stabilized and the protons are much less acidic. The small amount of exchange was attributed either to ylid formation and reprotonation or to a concerted reaction with no free ylid formed.

Tanaka, Dunning, and *Carter* [142] also studied this reaction and found that even at elevated temperatures in a sealed tube, the methyl hydrogen atoms of the tetramethylammonium ion did not exchange with solvent in dilute solution (0.6 M). Under these conditions the tetramethylammonium ion decomposes to give trimethylamine and methanol but no deuterium is incorporated into either the starting material or the products. Thus, the hydroxide ion in dilute solution does not behave as a proton abstracting agent. Although the tetramethylammonium ion will not undergo proton exchange with D_2O in dilute base, *Leitsch* [139] observed that exchange does occur if the reaction is carried out using a more concentrated solution (3 M) in a sealed tube at 130 °C and suggested that a repetitive exchange is useful for preparing the completely deuterated tetramethylammonium ion. Although some of the material decomposes to trimethylamine during the exchange, the other products were not identified.

Evidence concerning the acidity of the carbon atoms of the tetramethylammonium ion has been discussed by *Doering* and *Hoffman* [43] and by *Cram* [35]. Deuterium exchange in the tetramethylammonium ion in deuterated water at 83 °C proceeds with a rate constant of 9.4×10^{-10} sec^{-1}. In t-butyl alcohol-O-d at 50 °C, the rate constant is about 10^{-7} sec^{-1}. Thus the exchange rate is about 10^{-2} slower than the exchange rate in 3-phenyl-1-butene, which indicates that the acidity of the methyl protons in the quaternary salt is comparable to triphenylmethane

($pK_a = 33$) on the MSAD scale [35]. This value, which is slightly lower than the value for toluene ($pK_a = 35$), seems to be in conflict with the observation that benzylsodium does not cause deprotonation. However, it is suggested [35] that the lack of reaction with benzyl sodium may reflect an extremely slow proton transfer rather than an unfavorable free energy for the reaction.

In order to examine the last step proposed in the S_N2 mechanism for the decomposition of tetramethylammonium hydroxide, the decomposition of several tetramethylammonium alkoxides — in particular tetramethylammonium methoxide — were investigated [109]. The decomposition of mixtures of solid tetramethylammonium bromide with potassium methoxide, t-butoxide and triethyl carbinolate were carried out at 200 °C, in the absence of solvent. In all cases the predominant product was the ether, but a 20—25% yield of the corresponding alcohol was observed, indicating that some proton abstraction must have occurred. The steric requirement of the base did not affect the overall yield and the relative amount of ether and alcohol that was obtained.

Further insight into the course of the reaction was sought by a detailed examination of the decomposition of tetramethyl-1,1,1-d_3-ammonium methoxide in the absence of solvent [109]. If ylid formation occurs, deuterium scrambling will be observed and the resulting ether will be composed of CH_3OCH_3, CH_3OCDH_2, CH_3OCD_2H, and CH_3OCD_3. When the decomposition was carried out at several temperatures between 80 °C and 200 °C, deuterium scrambling in the ether was always observed and both methanol and methanol-d were obtained. The extent of scrambling was greater at low temperature than at high temperature suggesting that ylid formation is important at low temperature and that an S_N2 reaction may become important as the temperature is increased. It is also possible that proton abstraction occurs at high temperature but the ylid decomposes before intramolecular hydrogen scrambling can occur.

Analogous intramolecular hydrogen migration has been reported in methylenetrimethylphosphorane at 100 °C by *Schmidbaur* and *Tronich* [127,128,129] using nmr spectroscopy. A similar migration is also reported for trimethylsulfonium methylide. Since the methylenetriphenylphos-

$$
\begin{array}{ccc}
CH_3 & & ^-CH_2 \\
| & ^- & | \\
CH_3-P^+-CH_2 & \rightleftharpoons & CH_3-P^+-CH_3 \ \text{etc.} \\
| & & | \\
CH_3 & & CH_3
\end{array}
$$

phorane is stabilized by resonance, the nitrogen ylid might be expected to undergo exchange at a much lower temperature.

An alternative explanation for the scrambling of hydrogen involves an intermolecular exchange. Although this possibility cannot be eliminated, the decomposition of tetramethyl-1,1,1-d_3-ammonium methoxide in methanol solution led to a mixture of deuterated ethers which was similar to that observed in the dry salt. If an intermolecular exchange was occurring, more of the deuterium should have been incorporated into the solvent rather than equilibrating between ammonium ions.

Although some ylid formation undoubtedly occurs in these decompositions, it is not known whether the ether is formed by a carbenoid insertion into the O—H bond of the alcohol [14,82] or by reprotonation of the ylid followed by nucleophilic attack of methoxide. Although other ylid decomposition products were sought, no evidence for their existence was observed [109]. Several other reactions between alkoxides and the tetramethylammonium ion have been reported. The β-naphthoxide gives an 80% yield of the methylether on decomposition at 110 °C whereas the triphenylcarbinolate gives a 37% yield of the methylether and a 37% yield of the carbinol in boiling dioxane at 102 °C [162].

B. Amides

Amide ions have been used for many years to induce proton abstraction reactions and therefore might be expected to react with the tetramethylammonium ion in a similar way. In 1935, the decomposition of tetramethylammonium amide in liquid ammonia was initially reported by *Franklin* [44] to give trimethylamine and methylamine but no experimental details were given. Later, *Hazlehurst, Holliday*, and *Pass* [63] repeated *Franklin*'s experiment [44] and noted that a trace of ethylene was evolved. During later attempts to prepare solid tetramethylammonium amide, a violent explosion sometimes occurred as the last trace of ammonia was removed from the solid [107,112].

In a study of the decomposition of tetramethylammonium amide, *Musker* [107] found that the decomposition could be controlled and several products were detected when the ammonia solvent was removed at low temperature. The products isolated are those expected if the reaction proceeds *via* a nitrogen ylid, *e.g.*, trimethylamine, ethylene, polymethylene, and dimethylethylamine.

$$(CH_3)_3\overset{+}{N}-CH_3 + \overset{-}{N}H_2 \rightleftharpoons (CH_3)_3\overset{+}{N}-\overset{-}{C}H_2 + NH_3$$

$$\downarrow$$

$$(CH_3)_3N + CH_2{=}CH_2 + (CH_2)_n + (CH_3)_2N-CH_2-CH_3$$

Wittig, Heintzeler, and *Wetterling* [162] reported that the decomposition of tetramethylammonium piperidide gave 30% piperidine and 45% N-methylpiperidine. Thus, it appears that some proton abstraction occurs,

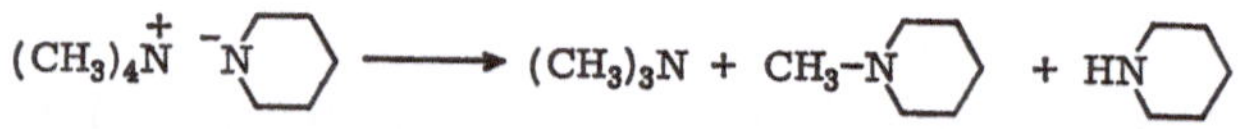

since piperidine is formed, but no ethylene, polymethylene or dimethylethylamine were observed.

C. Alkali Metals in Dioxane [52]

The reductive cleavage of quaternary ammonium salts to give a tertiary amine and a hydrocarbon by reaction with sodium amalgam in hydroxylic solvents is called the *Emde* degradation. However, saturated hydrocarbons are not cleaved under these conditions. *Grovenstein* suggested that the reason for this unreactivity is due to the fact that the sodium reacts with the alcohol much faster than with the ammonium salts. However, by operating in dioxane or dioxane-alcohol mixtures, the dealkylation reaction could be accomplished. The products of the decomposition of either tetramethylammonium chloride or bromide are methane, ethylene, trimethylamine, and dimethylethylamine.

The mechanism proposed for this reaction was written as follows:

$$(CH_3)_3\overset{+}{N}-CH_3 + 2\,Na \;\rightarrow\; CH_3Na + (CH_3)_3N + Na^+ \tag{1}$$

$$(CH_3)_3\overset{+}{N}-CH_3 + CH_3Na \;\rightarrow\; (CH_3)_3\overset{+}{N}-\overset{-}{C}H_2 + CH_4 + Na^+ \tag{2}$$

$$(CH_3)_3\overset{+}{N}-\overset{-}{C}H_2 + (CH_3)_3N-CH_3 \;\rightarrow\; (CH_3)_3\overset{+}{N}-CH_2-CH_3 + (CH_3)N \tag{3}$$

$$(CH_3)_3\overset{+}{N}-CH_2-CH_3 + B^- \;\rightarrow\; (CH_3)_3N + CH_2-CH_2 + BH \tag{4}$$

The route to dimethylethylamine was not discussed, but probably involves a *Stevens* rearrangement of the ylid or methyl group displacement subsequent to step *(3)*. Alkylation of the ylid by the tetramethylammonium ion followed by a β elimination of ethyltrimethylammonium ion (Step *4*) was suggested by *Wittig* and *Krauss* [179] to account for the formation of ethylene [171].

In the course of this reaction no hydrogen gas was observed, although under the experimental conditions it could not have been detected. If hydrogen gas is also evolved along with methane, then ylid formation may have occurred by the reaction of an electron with the tetramethylammonium ion.

$$(CH_3)_3\overset{+}{N}-CH_3 + e^- \;\rightarrow\; (CH_3)_3\overset{+}{N}-\overset{-}{C}H_2 + H\cdot$$

$$2\,H\cdot \;\rightarrow\; H_2$$

Actually a two electron reduction leading to the hydride ion can also be postulated even if hydrogen gas is not observed, since the hydride ion probably would react with the tetramethylammonium ion

$$(CH_3)_3\overset{+}{N}-CH_3 + 2\,e^- \;\rightarrow\; (CH_3)_3\overset{+}{N}-\overset{-}{C}H_2 + H^-$$

$$H^- + (CH_3)_4\overset{+}{N} \;\rightarrow\; CH_4 + (CH_3)_3N$$

to give methane under the conditions of the reaction. For example, lithium aluminium hydride demethylates quaternary ammonium salts in ether to give methane [33].

Substituents larger than the methyl group were also examined in this study [52] and both olefin formation and reductive cleavage was observed. It is worth noting that the distribution of products was quite dependent on the anion.

Radical intermediates in this reaction were not proposed; however, in light of recent studies of radical intermediates [98] in various ylid decomposition processes, this possibility should be examined. (Sect. VI. C.)

$$
\begin{array}{c}
\underset{\displaystyle \underset{CH_3}{|}}{\overset{\displaystyle \overset{CH_3}{|}}{CH_3-N^+-CH_2}} \quad \rightarrow \quad \left[\underset{\displaystyle \underset{CH_3}{|}}{\overset{\displaystyle \overset{CH_3}{|}}{CH_3 \cdot \quad \cdot N - \ddot{C}H_2}} \right] \\[3ex]
\updownarrow \\[2ex]
\underset{\displaystyle \underset{\ddot{}}{|}}{\overset{\displaystyle \overset{CH_3}{|}}{CH_3-N-CH_2CH_3}} \quad \leftarrow \quad \left[\underset{\displaystyle \underset{CH_3}{|}}{\overset{\displaystyle \overset{CH_3}{|}}{CH_3 \cdot \quad :N - \dot{C}H_2}} \right] \rightarrow CH_4 + (CH_3)_3N
\end{array}
$$

D. Alkali Metals in Liquid Ammonia

The reaction of metallic sodium and potassium in liquid ammonia with tetraalkylammonium salts was initially studied by *Thompson* and *Cundall* [143] and extended by *Grovenstein* and coworkers [53,54,55] and by *Hazlehurst, Holliday*, and *Pass* [63].

During an attempt to prepare the tetramethylammonium radical, *Thompson* and *Cundall* [143] reported that the reaction of tetramethylammonium iodide with potassium in liquid ammonia gave trimethylamine, ethane, and potassium iodide, but no methane.

$$(CH_3)_4\overset{+\,-}{N}I + K \xrightarrow[NH_3]{} CH_3-CH_3 + (CH_3)_3N + KI$$

Grovenstein and *Stevenson* [53] reported that the reaction of the bromide salt with sodium in liquid ammonia at $-33\,°C$ gives only trimethylamine and methane with a trace of ethylene. No hydrogen gas or dimethylethylamine were reported but they might have escaped detection under the experimental conditions employed. Based on product analysis, the following mechanisms were suggested in which either methyl radicals or methyl carbanions are formed.

$$(CH_3)_3\overset{+}{N}-CH_3 + e^- \rightarrow (CH_3)_3N + CH_3 \cdot$$
$$CH_3 \cdot + e^- \rightarrow CH_3^-$$

or

$$(CH_3)_3N-CH_3 + 2\,e^- \rightarrow (CH_3)_3N + CH_3^-$$
$$CH_3^- + NH_3 \rightarrow CH_4 + NH_2^-$$

In the reactions of sodium in liquid ammonia with quaternary ammonium salts with different alkyl groups the methyl group is always cleaved first (except for the *t*-butyl group). This behavior is similar to LAH reductions [33] and to thermal decompositions of halide salts [92].

Hazelhurst, Holliday, and *Pass* [63] also studied this reaction, but their conditions were slightly different — potassium in liquid ammonia at −78 °C on a vacuum line. The main difference between their results and those reported by *Grovenstein* is that hydrogen gas is observed along with traces of ethane and ethylene but no dimethylethylamine. The other products were the same.

The reactions proposed by these workers are similar to those proposed by *Grovenstein* [53], except that they attribute some of the hydrogen gas to ammonia decomposition and the ethane to a recombination of methyl radicals formed in the reaction.

$$NH_3 + e^- \rightarrow NH_2^- + \tfrac{1}{2}H_2$$

$$2\,CH_3 \cdot \rightarrow C_2H_6$$

It is worth noting that the yield of hydrogen is greater when the bromide and iodide salts are used rather than the chloride. To explain the formation of ethylene, an ylid was postulated which was formed by attack of an electron on the tetramethylammonium ion. It was assumed [63] that the ylid was stable and did not decompose to give ethylene nor rearrange to give dimethylamine — an assumption which was later shown to be incorrect [179].

Therefore an ylid mechanism satisfactorily accounts for the ethylene, hydrogen, and amide ion which is produced.

$$(CH_3)_3\overset{+}{N}-CH_3 + e^- \rightarrow (CH_3)_3\overset{+}{N}-\overset{-}{C}H_2 + \tfrac{1}{2}H_2$$

$$(CH_3)_3\overset{+}{N}-\overset{-}{C}H_2 \rightarrow CH_2{=}CH_2,\ (CH_2)_n,\ CH_3-N-CH_2-CH_3,\ (CH_3)_3N$$
$$\underset{CH_3}{|}$$

$$(CH_3)_3\overset{+}{N}-\overset{-}{C}H_2 + NH_3 \rightarrow (CH_3)_3\overset{+}{N}-CH_3 + \overset{-}{N}H_2$$

It is unfortunate that no polyethylene or dimethylethylamine was observed since this would have helped to substantiate the presence of the ylid.

The formation of methane was explained [53] by the generation of a methyl carbanion which reacts with solvent. The reaction of the methyl carbanion with the tetramethylammonium ion could account for the traces of ethane observed *Hazlehurst, Holliday,*and *Pass* [63]. Alternatively recombination of methyl radicals or reductive cleavage of the trimethyl-ethylammonium ion (formed by alkylation of the ylid by the tetra-methylammonium ion) could also account for the ethane.

Several alkyltrimethylammonium salts were also examined in this study [53]. The predominant reaction is reductive cleavage to give the alkane and tertiary amine. In the discussion of the mechanism of the reaction, it was suggested [53] that methyl and other primary alkyl groups cleave from nitrogen as carbanions while secondary and tertiary alkyl groups cleave as free radicals (see Sect. VI.C. for further discussion of radical cleavage reactions).

IV. Reactions of the Tetramethylammonium Ion with Weak Bases

The thermal decomposition of a variety of tetramethylammonium salts of weak bases has been studied and the major product is generally the methylated anion [92]. The temperature required to decompose the bromide and chloride salts to trimethylamine and the methyl halide is near 360 °C and the products recombine on cooling. The fluoride salt decomposes at a much lower temperature (180 °C) and the products do not recombine [92]. Decomposition of the borohydride salt at 225 °C gives trimethylamine borane and methane [11].

$$(CH_3)_4 \overset{+}{N} BH_4^- \rightarrow CH_4 + (CH_3)_3NBH_3$$

Robb and *Westbrook* [119] heated a variety of carboxylate salts (180—250 °C) to give the methyl esters in high yield. The decomposition of the salts of other weak bases (NO_3^-, SO_4^{2-}, $C_2O_4^{2-}$, citrate, malonate, etc.) was also studied but complex mixtures of products were obtained which were not fully characterized.

Recently *Wilson* and *Joule* [158] studied the demethylation of a variety of quaternary ammonium acetates in aprotic solvents. Earlier *Lawson* and *Collie* [92] showed that the decomposition of solid tetramethyl-ammonium acetate at 180—200 °C gives the methyl ester in good yield. However, the use of an aprotic solvent significantly lowers the temperature (60—140 °C) required for decomposition. This reaction is most useful for the demethylation of aromatic quaternary ammonium salts that are soluble in benzene or benzene-chloroform mixtures. Its application to aliphatic quaternary ammonium salts requires a longer time, although N,N-dimethylpiperidinium acetate was demethylated at 100 °C in xylene-acetonitrile solvent to give an 88% yield of N-methylpiperidine. The addition of 1% methanol resulted in no reaction. The reaction is faster in solvents with low dielectric constant. An explanation for the superiority of non-polar solvents is attributed to their poor solvation of the starting materials relative to the less ionic transition state for demethylation.

$$B^- + CH_3NR_3 \rightarrow \left[B^{\delta^-} \ldots \ldots \overset{H}{\underset{H}{\overset{\diagup}{C}}} \diagdown^H \ldots \ldots \delta^+ \overset{R}{\underset{R}{N}} \diagdown R \right] \rightarrow BCH_3 + NR_3$$

The reaction of *thiophenoxide* with simple quaternary ammonium salts was reported by *Shamma, Deno,* and *Remar* [134] in 1966. They reported that the dealkylation of triethylmethylammonium thiophenoxide could be achieved in ~ 95% yield by heating the salt to 80 °C for 19 hours in solvents such as 2-butanone or acetonitrile or by heating the dry salt for 2 hours. Demethylation was far more prevalent than deethylation by a factor of 3 and no ethylene was produced.

$$(C_2H_5)_3\overset{+}{N}-CH_3 \; \overset{-}{S}\phi \rightarrow CH_3S\phi + (C_2H_5)_3N$$

However, *McKenna* and coworkers [103] have reported that in heterocyclic N-ethyl-N-methyl quaternary ammonium salts, the extent of demethyl and deethylation with thiophenoxide in triethyleneglycol depends on the orientation of the group and not necessarily on its size. Since no ethylene is produced, the reaction probably proceeds by a pure S_N2 reaction without hydrogen scrambling and therefore without ylid formation.

To test this conclusion, the decomposition of tetramethyl-1,1,1-d_3-ammonium thiophenoxide was studied [111]. The decomposition of the dry salt was carried out on a vacuum line giving a 3 to 1 mixture of pure methylphenyl sulfide and trideuteromethylphenylsulfide without any proton scrambling.

$$CH_3-\overset{\overset{\displaystyle CH_3}{\displaystyle |}}{\underset{\underset{\displaystyle CH_3}{\displaystyle |}}{N^+}}-CD_3 \; \overset{-}{S}\phi \rightarrow CH_3S\phi + CD_3S\phi$$

Since thiophenoxide apparently reacts by a pure S_N2 process, the reaction of other good nucleophiles with the tetramethylammonium ion probably proceeds in a similar way.

Several other reagents have been used to affect demethylation, such as lithium iodide [70], ethanolamine [74], and morpholine [12]. With these reagents, simple displacement reactions are observed.

V. Ylids from Phenyl Substituted Quaternary Ammonium Salts

Under conditions of the *Hofmann* degradation (heating an aqueous solution of the hydroxide until the water is evaporated and decomposition occurs), trimethylphenylammonium hydroxide *(5)* demethylates exclusively to give N,N-dimethylaniline and methanol [27].

$$
\begin{array}{ccc}
\underset{\underset{\displaystyle 5}{}}{\underset{\displaystyle \bigcirc}{\overset{\displaystyle CH_3}{\underset{|+}{CH_3-N-CH_3}}}} & \xrightarrow{\ OH^-\ } & \underset{\displaystyle \bigcirc}{\overset{\displaystyle }{CH_3-N-CH_3}} \quad + \ CH_3OH
\end{array}
$$

When *5* is treated with amide ion in liquid ammonia, N,N-dimethylaniline, methylamine, trimethylamine, and aniline are formed [120]. No evidence for proton abstraction from the methyl group was noted, but

$$5 \ + NH_2^- \rightarrow C_6H_5N(CH_3)_2 + C_6H_5NH_2 + (CH_3)_3N + CH_3NH_2$$

proton abstraction from the ring to give benzyne was observed [120]. When *5* is treated with phenyllithium in THF, reaction occurs rapidly with the formation of benzene, trimethylamine, N,N-dimethylaniline, and N-methyl-N-ethylaniline [157]. Biphenyl and small amounts of ethylene and acetaldehyde were also observed.

The formation of N-methyl-N-ethylaniline and benzene can be explained by a proton abstraction followed by a *Stevens* rearrangement of the methylene ylid [157]. An immonium ion pair

$$
5 \ + C_6H_5Li \xrightarrow{\ THF\ } C_6H_5-\overset{\overset{\displaystyle CH_2Li}{|}}{\underset{\underset{\displaystyle CH_3}{|}}{N^+}}-CH_3 + C_6H_6
$$

$$
\underset{\underset{\displaystyle CH_3}{|}}{\overset{\overset{\displaystyle CH_2-CH_3}{|}}{C_6H_5-N:}} \longleftarrow \left[\underset{\underset{\displaystyle CH_3}{|}}{\overset{\overset{\displaystyle CH_2}{\|}}{C_6H_5-N^+}}-CH_3^-\right] \longleftarrow \quad \longrightarrow [CH_2] + C_6H_5N(CH_3)_2
$$

was suggested as the intermediate leading to N-methyl-N-ethylaniline. Since no toluene was observed, the N,N-dimethylaniline could not have resulted from a displacement reaction and must be formed by decomposition of the ylid. Dimerization of the carbene to give ethylene is unlikely, but if the carbene inserts into the C—N bond of 5, ethylene could result from an α'-β elimination (Sect. VII) of the dimethylethylphenylammonium ion by phenyllithium [157].

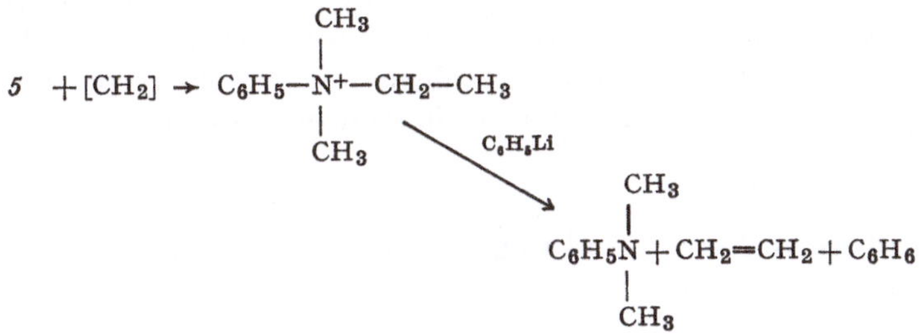

$$5 \;\; +[CH_2] \;\rightarrow\; C_6H_5{-}\overset{\overset{\displaystyle CH_3}{\displaystyle |}}{\underset{\underset{\displaystyle CH_3}{\displaystyle |}}{N^+}}{-}CH_2{-}CH_3$$

$$\xrightarrow{\;C_6H_5Li\;} \quad C_6H_5\overset{\overset{\displaystyle CH_3}{\displaystyle |}}{\underset{\underset{\displaystyle CH_3}{\displaystyle |}}{N}} + CH_2{=}CH_2 + C_6H_6$$

Alternatively, the N-methyl-N-ethylaniline could also be accounted for by carbene insertion into the C—H bonds of dimethylaniline [45]. However, carbon-14 labeling experiments have shown that the ethylene arises from the methyl group and also from the cleavage of the THF by phenyllithium [157].

Since considerable trimethylamine was observed, the fate of the phenyl group was examined. It was determined by carbon-14 labeling, that biphenyl results from the reaction of phenyllithium with 5. However, it is most unlikely that the phenyl group can be displaced by phenyllithium in an S_N2 reaction. Therefore, it was suggested [157] that proton abstraction occurs at the *ortho* carbon of the benzene ring to give *benzyne* which subsequently reacts with phenyllithium to give biphenyl.

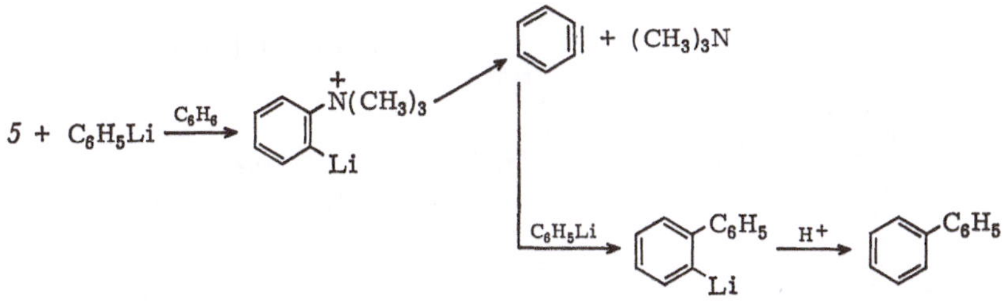

To test for the presence of a benzyne intermediate, the reaction was carried out in the presence of lithium thiophenoxide. The reaction proceeded much slower and diphenylsulfide was found as one of the products of the reaction. The thiophenoxide salt must have reacted with benzyne to form this product for it does not react with 5 in the absence of phenyllithium.

Recently methyltriphenylammonium tetrafluoroborate was prepared from triphenylamine and trimethyloxonium fluoroborate. Demethylation was affected by butyllithium in hexane or methylene chloride and with phenyllithium in benzene or ether. No evidence for biphenyl or diphenylmethylamine was noted and only triphenylamine was characterized. In demethylation reactions with potassium methoxide in methanol-O-d, no exchange of methyl hydrogen for deuterium was observed. Thus, no proton abstraction processes occurred in these decompositions [114].

VI. Ylids from Benzyl and Benzhydryl Substituted Quaternary Ammonium Salts

In recent years the reactions of basic reagents with quaternary ammonium salts containing the benzyl and benzhydryl groupings have received extensive study. It is useful to examine these reactions for the existence of ylids and to note the variety of ways in which they react.

This section will not include a complete discussion of benzyl substituted quaternary ammonium salts since several reviews of these compounds are currently available [10,30,183].

A. Benzyl Substitution

Historically, *Hughes* and *Ingold* [71] reported in 1933, that the *Hofmann* degradation of benzyltrimethylammonium ion 6 gives mainly displacement products.

The methanol was determined qualitatively and trace amounts of benzaldehyde and dibenzylether were also detected. No evidence for any rearrangement products was found. From these observations the reaction seems to be similar to the decomposition of tetramethylammonium alkoxides and hydroxide, but since no proton scrambling experiments have been performed, it is difficult to ascertain whether proton abstraction occurs in this reaction. It is worth noting also that no dimethyl ether was reported, although this product could have escaped detection under the conditions of the reaction.

On treatment with sodium amide in liquid ammonia, 6 undergoes a *Sommelet* rearrangement [138] (*ortho*-rearrangement) [81] to give 7 in 80—90% yield. *Kantor* and *Hauser* [81] initially postulated an ylid mechanism

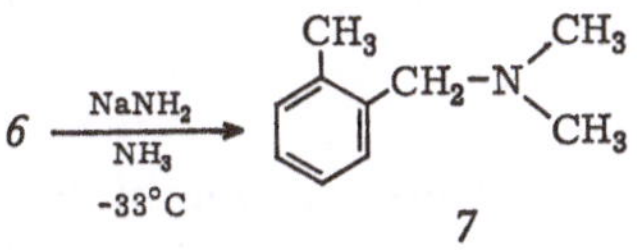

$$6 \xrightarrow[\substack{NaNH_2 \\ NH_3 \\ -33°C}]{} \quad \text{7}$$

to explain the formation of the product. Subsequently, carbon-14 experiments [79,80] proved that the *ortho*-methyl group in 7 was indeed the benzyl carbon atom of 6 and it was suggested that the methylene ylid 6_m undergoes a nucleophilic attack at the *ortho*-carbon of the benzene ring (6_{em}),

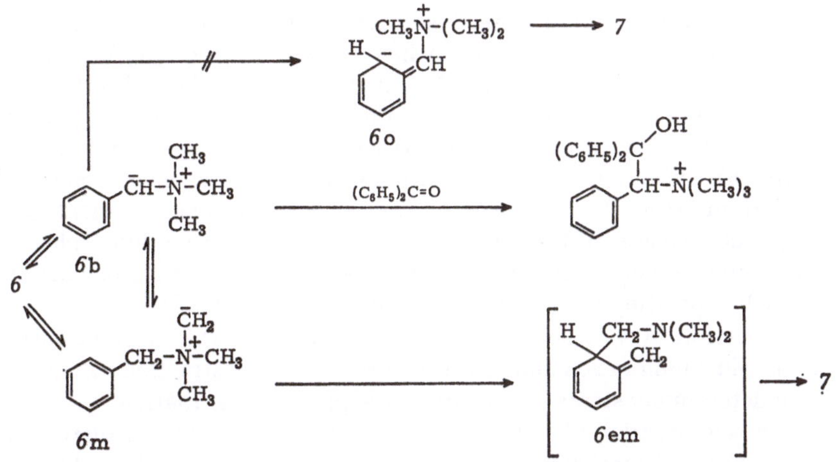

rather than isomerizing to $6o$ which undergoes a nucleophilic attack at the $N^+\text{—}CH_3$ group. Later *Puterbaugh* and *Hauser* [118] attempted to trap the intermediate methylene ylid $6m$ by condensation with benzophenone at -80 °C, but they only succeeded in isolating the adduct of the more stable benzyl ylid $6b$. Since the benzyl ylid $6b$ is known to give 7 on warming to -33 °C, an inter- or intramolecular ylid equilibrium was proposed to give *6m* which then reacts repidly to give *7*. Using deuterium labeling, *Pine* [116] observed the equilibration of protons between the benzyl and methyl groups in this reaction.

The reaction of *6* with phenyllithium was reported [164] to give the *Stevens* rearrangement product *8* as the only rearrangement product along with a small amount of 1,1,2-triphenylethane. It was suggested that *8* is formed *via* the benzyl ylid intermediate *6b*. In a later study,

in which the effects of various organolithium reagents were studied, *Lepley* and *Becker* [95] observed that both *7* and *8* were formed in the reaction of *6* with phenyllithium and that *8* was formed in the greater amount.

However, when *n*-butyllithium was used, twice as much *7* as *8* was observed. With sec-butyllithium a similar ratio of products was noted, but the overall yield of rearranged amines was lower. It was noted [95] that the anion may influence this reaction just as it does in the reaction of tetraalkylammonium salts. For example, the iodide of *6* was much less reactive than the chloride when treated with n-butyllithium in pentane, but differences in solubility may account for these observations.

In general it was found [95] that the overall yield of rearrangement products increases with increasing basicity of the organolithium reagent with the exception of *sec*-butyllithium where steric effects may be important. However, the ratio of *7/8* using the various organolithium reagents decreases in the series n-butyl > sec-butyl > methyl > phenyl. This order is not in line with differences in basicity and other factors must be operating to account for the observed trend.

Therefore it was suggested [95] that rearrangement must be faster than ylid equilibration and that the initial proton abstraction step is the rate determining step which controls the product distribution.

In this situation of a slow methyl ylid formation and fast rearrangement, the ylid mechanism is not distinct from a concerted mechanism and perhaps is less favorable. The following mechanism, which does not involve a free ylid, was then proposed for the formation of *7* [95].

However, it was suggested that *8* was formed from the benzyl ylid since its presence had been demonstrated by the trapping procedure [95].

The presence of an exomethylene cyclohexadiene *(6em)* in the reaction of *6* with butyllithium in hexane at 25 °C was substantiated by *Pine* and *Sanchez* [117], who isolated the following compound, presumably formed by the attack of butyllithium on the exomethylene group.

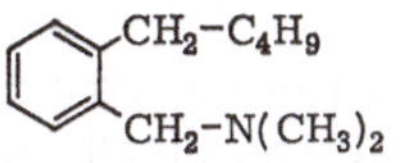

CH2–C4H9
CH2–N(CH3)2

In a similar study, *Klein, Van Eenam,* and *Hauser* [83] reported the reaction of *6* with butyllithium in ether-hexane at 0–5 °C. Under these conditions, several products, *9–11*, which were not detected in previous work, were obtained in small amounts. The following sequence of reactions was proposed to account for the products.

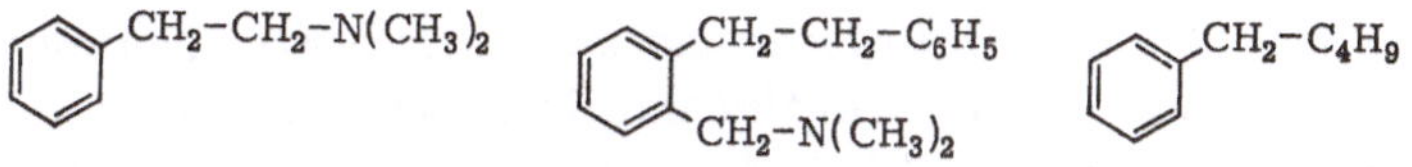

CH2–CH2–N(CH3)2
CH2–CH2–C6H5
CH2–N(CH3)2
CH2–C4H9

9 *10* *11*

6
a. n-C4H9Li
Li CH3
C6H5–CH–N±CH3
CH3
8
b. n-C4H9Li
n-C4H9Li
CH2Li
C6H5CH2–N±CH3
CH3
9
Li CH2Li
C6H5–CH–N±CH3
CH3
CH2Li
CH2–N(CH3)2
6
10
12
n-C4H9Li
H2O
7

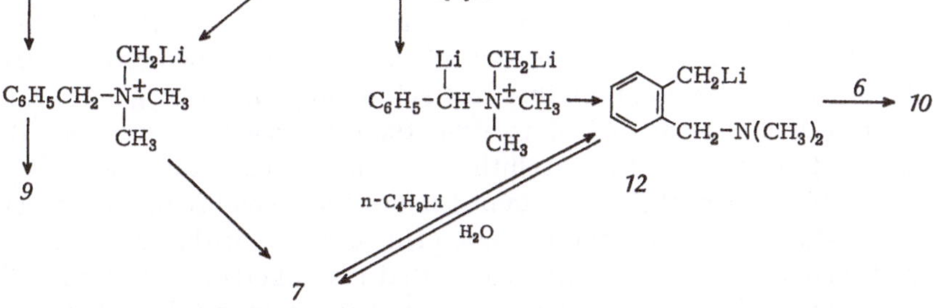

This mechanism differs from the preceeding mechanism primarily in that the lithium ion is used to stabilize the ylid and a dilithium reagent is thought to be the intermediate leading to *7* and *10*. The existence of dilithium reagents has been noted in the reactions of the tetramethylammonium ion with excess phenyllithium in ether [170] and it is quite possible that they can be formed here.

The existence of a dilithium reagent was suggested partly because *12* could be trapped with benzophenone.

$$12 \; + \; (C_6H_5)_2C{=}O \; \longrightarrow \; \underset{CH_2-N(CH_3)_2}{\overset{CH_2-\overset{\overset{\textstyle OH}{|}}{C}-(C_6H_5)_2}{\bigcirc}}$$

However the isolation of this compound cannot be used as definite evidence for *12*, since *7* can also be metallated by butyllithium to give *13* under the reaction conditions. However, when *12* is used as the base in the reaction, the major product is *8* rather than *7* [83]. This would be expected since *12* is a much weaker base than butyllithium and would not be as capable of abstracting a proton from the N^+–CH_3 group to give the required dilithium reagent.

The existence of the methyl ylid under these experimental conditions receives support from the formation of *9* which was characterized by the g.c. peak enhancement technique [83]. The precursor of *9* must have been *6m* which undergoes a *Stevens* rearrangement with benzyl migration to give *9*. The absence of *9* under *Lepley* and *Becker*'s conditions [95] was one of the reasons that they ruled out the existence of the free ylid. It is unfortunate that neither the methyl ylid nor its lithio derivative could be trapped. It is also worth noting that in this formulation of the ylid, the anion is not included as it is throughout the discussion of trimethyl-ammoniummethylide.

Recently *Klein* and *Hauser* [84] used the *dimethylsulfinyl carbanion* to convert *6* to *7* in 81—85% yield. This method is far more convenient for the synthesis of *7* than the sodium amide method, for the yield is comparable and the reaction can be carried out at room temperature. *Lepley* and *Brodof* [97] reported another method of preparing *7* in good yield. Their technique consisted of treating the ether soluble p-(tert-butyl)-phenoxide salt of *6* with butyllithium at low temperature (0—15 °C). An 88—92% yield of *7* was obtained under these conditions. At higher temperatures, however, the amount of *8* increases with decreasing *7*. In this later work it was suggested [97] that two different paths lead to *7* and *8*. The benzyl ylid is the precursor to the *Stevens* product *8* as suggested earlier. However, an ion pair intermediate is the precursor to *7*.

$$6 \; \longrightarrow \; \left[\bigcirc{-}\bar{C}H_2 \;\; \overset{\overset{\textstyle CH_2}{\overset{\textstyle \|}{}}}{\underset{\underset{\textstyle CH_3}{|}}{N}}{-}CH_3 \right] \; \longrightarrow \; 7$$

This suggestion receives support from the fact that the yield of 7 increases as the solvent polarity is changed from ethers to DMSO in keeping with a greater charge separation in the ion pair than in the ylid.

Pine [115] reported the first example of a *para-Sommelet-Hauser* rearrangement by using α-phenylneopentylammonium salts. The yield of *para*-rearrangement product varies from 0 to 10%, depending on the anion and the solvent.

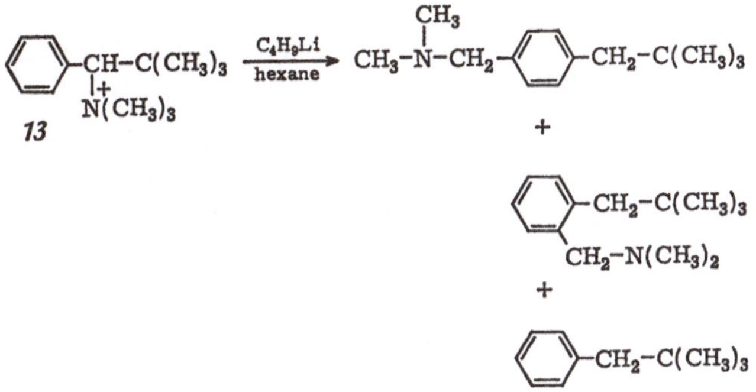

An internal ion pair mechanism was proposed to account for the *para*-rearrangement product [115] but since the yield is rather small, the ion-pair could dissociate and then recombine in hexane solvent rather than recombine within the solvent cage.

The transition state leading to an ion-pair intermediate may be the free ylid 6_m; however, proton abstraction from a methyl group with concomitant cleavage of the $N^+-CH_2C_6H_5$ bond gives the same ion pair without the existence of a free ylid.

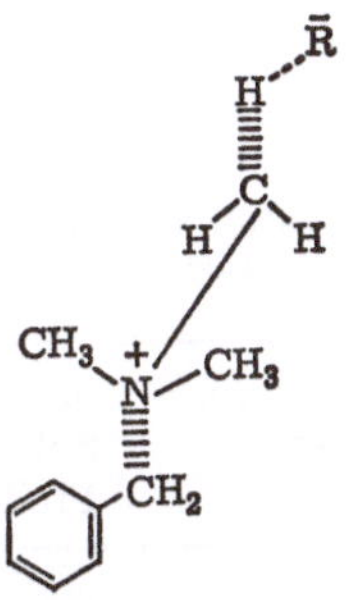

The difference between these two mechanisms lies in the relative amount of C—H and N—C bond breaking in the transition state [24]. Further discussion on the formation of ion pairs will be presented later. (Sect. VI.D.)

B. Dibenzyl Substitution

In the reaction of dibenzyldimethylammonium iodide *14* with a) sodium amide in liquid ammonia, b) phenyllithium in ether, and c) fused sodium amide at 145 °C [145], the relative amount of the *Stevens* product, *15*, [166] to *ortho*-substitution product, *16*, increases with increasing temperature [59].

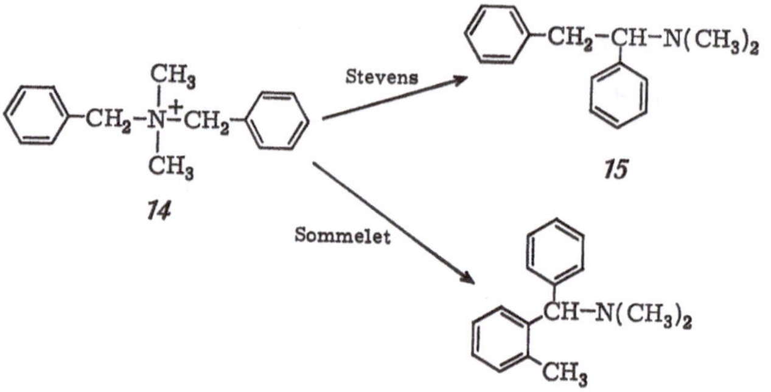

An extensive study of the temperature dependence on product distribution was carried out by *Wittig* and *Streib* [169] on a similar compound, *17*.

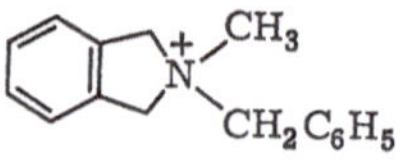

Variation in ortho-Substitution and Stevens Rearrangement Products from 17 [169]

	Temp. [°C]	ortho-Substitution Product	Stevens Product
NaNH₂ (NH₃)	−33	87	—
φLi (Et₂O)	20—30	67	—
NaOEt (EtOH)	80	—	44
φLi (Bu₂O)	120	—	41
Hofmann dec.	180	—	33

By an analysis of the ratio of the rate constants for the two competing reactions of the dibenzyldimethylammonium ion, *Zimmerman* [183] concludes that the entropy of activation for the *ortho* rearrangement must be more negative than for the *Stevens* rearrangement. Since a five-membered activated complex is required for the *ortho*-rearrangement, whereas only a three-membered activated complex is necessary for the *Stevens* rearrangement, this result is expected.

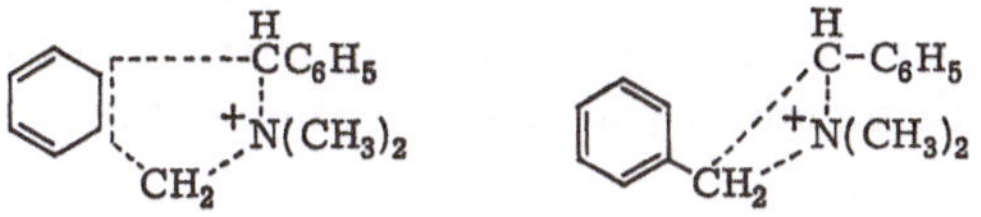

By invoking a quantum mechanical argument for the energy of the activated complex, *Zimmerman* [183] accounted for the fact that the enthalpy of activation of the *Stevens* rearrangement must be greater than the *ortho* rearrangement. In the model proposed by *Zimmerman*, the activated complex is described as the merging of a benzyl carbanion and an immonium cation. Using the LCAO coefficients for the non-bonding MO in the benzyl carbanion and the anti-bonding MO of the immonium cation, the energy of the transition state leading to the *Stevens* product was calculated to be greater than the transition state leading to the *ortho* rearrangement.

When *14* is treated with butyllithium in hexane [96] the rearrangement product *18* is found in appreciable yield in addition to *15* and *16*. This product results from an *ortho*-rearrangement involving the migration

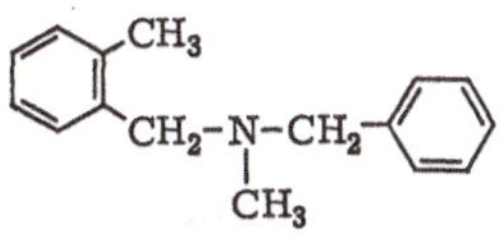

18

of a methyl ylid rather than a benzyl ylid. Although the rearrangement products are the most interesting compounds obtained, the major products are N,N-dimethylbenzylamine and *n*-pentylbenzene. These products are found in equivalent amounts and probably result from a displacement reaction at the benzylcarbon atom by butyllithium. Although displacement reactions by organolithium reagents are not generally observed in high yields, the isolation of 1,1,2-triphenylethylene in the decomposition of benzyltrimethylammonium by phenyllithium is also attributed to an initial displacement at the benzyl carbon by the phenyl anion [164].

Trans-stilbene is found as one of the minor products of the reaction [83]. Although a cleavage of the benzyl ylid to give phenyl carbene followed by dimerization could account for the stilbene, this reaction is considered to be less likely than a stepwise process which initially involves a displacement on *9* by the benzyl ylid followed by an α'-β elimination reaction [83].

It was suggested by *Lepley* and *Guimanini* [96] that the intimate immonium ion pairs *19* and *20* are the precursors to the rearrangement products rather than ylids or ylid anions because of the isolation of traces

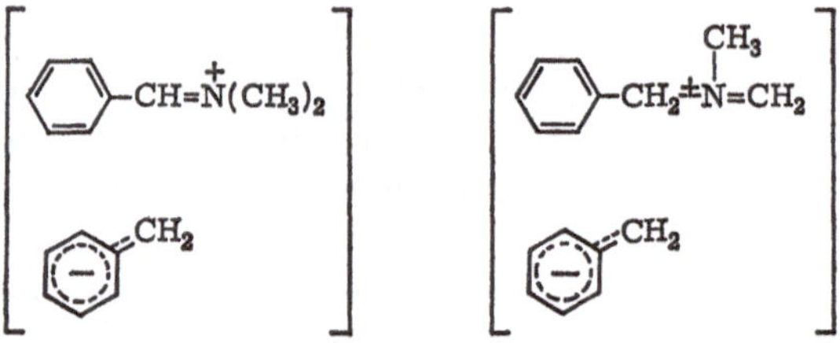

of toluene from the reaction mixture. *19* seems to be a reasonable suggestion, since it is formed from the benzyl ylid and both cationic and anionic partners have an independent stability; however, the presence of *20* as a precursor to *17* seems to be less likely since it would have to be formed by deprotonation at the methyl group. Perhaps, the ylid *21*, stabilized by a lithium ion, would have less charge separation and would be the predominant species in this solvent although dilithium derivatives are also possible [83].

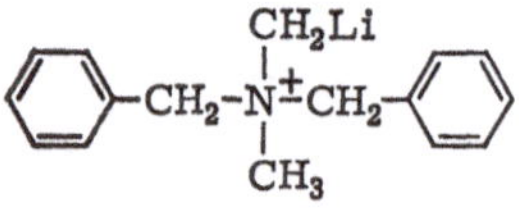

The lithiated intermediates could then rearrange to products *via* ion pairs or concerted intramolecular displacement reactions. Perhaps the major objection to invoking ylid or lithiated ylid intermediates is that more products should have been detected than were actually observed.

C. Benzyl Substituted Anilines

An extensive study of the decomposition of the N,N-dimethylbenzyl-anilinium ion 22 was recently carried out by *Lepley* and coworkers [93,99] As expected, displacement reactions occur exclusively with hydroxide and alkoxide, an *ortho*-rearrangement occurs with amide, and a variety of products are formed with various organolithium reagents. The products found when butyllithium was employed are listed below.

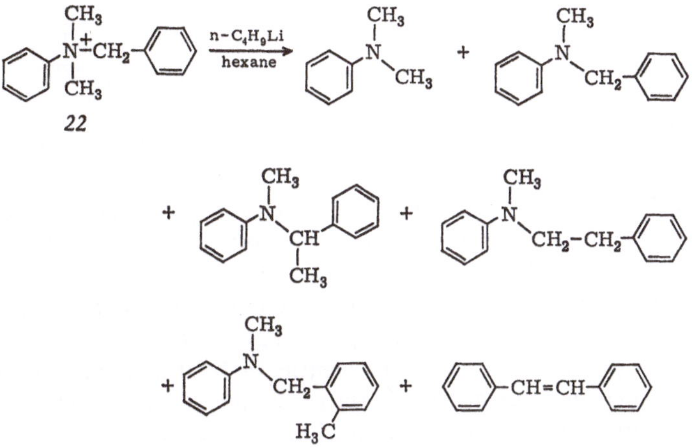

The yield and ratio of the two *Stevens* rearrangement products are dependent on both the base and the solvent; however, the *ortho*-rearrangement product was only detected when butyllithium was used in hexane. Variation of the halide ion had only a small effect on the ratio of rearrangement products. Several mechanisms were considered [93] as routes to the formation of the rearrangement products, 1) ion-pairs, 2) cationic rearrangement, 3) carbenoid, 4) dimetallated intermediates, 5) free-radical, 6) predissociation of the ammonium salt and, 7) sigmatropic shift.

The radical mechanism was considered as one of the best explanations for this rearrangement on the basis of a related study of the reaction of benzyne with N,N-dimethylbenzylamine in which a similar ylid is formed. (See Sect. IX.)

The cleavage products (N,N-dimethylaniline, N-methyl-N-benzyl-aniline and n-pentylbenzene) can be formed by simple displacement reactions on the starting material by the butyllithium. A free-radical cleavage was also considered as a possible, but less likely, alternative.

D. Ion-Pair Formation from Ylids

The suggestion of ylids decomposing to ion pair intermediates has been proposed in similar rearrangements [140,141,144]. *Thomson* and *Stevens* [144] suggested an ion-pair intermediate in the rearrangement of phenacyl-benzyldimethylammonium bromide *23* with sodium in liquid ammonia.

$$
\underset{23}{\overset{O}{\overset{\|}{C_6H_5-C}}-CH_2-\overset{CH_2C_6H_5}{\overset{|}{N^+}}-(CH_3)_2} \rightarrow \overset{O}{\overset{\|}{C_6H_5C}}-\overset{CH_2C_6H_5}{\overset{|}{\overset{-}{CH}}-N^+}-(CH_3)_2 \rightarrow \overset{O}{\overset{\|}{C_6H_5-C}}-CH=\overset{\overline{CH_2C_6H_5}}{N^+}-(CH_3)_2
$$

$$
\downarrow
$$

$$
\overset{O}{\overset{\|}{C_6H_5-C}}-\overset{CH_2C_6H_5}{\overset{|}{CH}}-N(CH_3)_2
$$

In this case the reaction is carried out in a polar solvent and both anion and cation partners are stabilized by resonance. Later it was shown [17] that the migrating group retains its configuration and therefore a tight ion-pair would have to be formed in order to prevent racemization. An ion-pair intermediate containing a benzyl carbanion was supported by the observation that electron accepting substituents in the benzene ring of the benzyl group accelerate the reaction whereas electron releasing groups retard the reaction. No appreciable effect on the reaction rate is noted when substituents are placed on the benzene ring of the phenacyl group [144]. Recently, however, *Schöllkopf* and *Ludwig* [131] have suggested that a radical pair intermediate rather than an ion-pair may be involved in this rearrangement.

$$
\overset{O}{\overset{\|}{C_6H_5C}}-\overset{CH_2C_6H_5}{\overset{|}{CH}}-\overset{|}{\underset{CH_3}{\overset{+}{N}}}-CH_3 \rightarrow
\left[
\begin{array}{c}
\overset{O}{\overset{\|}{C_6H_5-C}}-CH-\overset{..}{N}-(CH_3)_2 \\
\quad - \quad + \\
\\
\overset{O}{\overset{\|}{C_6H_5-C}}-\overset{.}{CH}-\overset{..}{N}-(CH_3)_2 \cdot CH_2C_6H_5 \\
\\
\overset{O\cdot}{\overset{|}{C_6H_5-C}}=CH-\overset{..}{N}-(CH_3)_2
\end{array}
\right] \rightarrow
$$

$$
\overset{O}{\overset{\|}{C_6H_5-C}}-\overset{CH_2-C_6H_5}{\overset{|}{CH}}-N(CH_3)_2
$$

This view was based on a kinetic analysis of the *Meisenheimer* rearrangement of the related amine oxide where a radical dissociation-recombination mechanism was proposed [131].

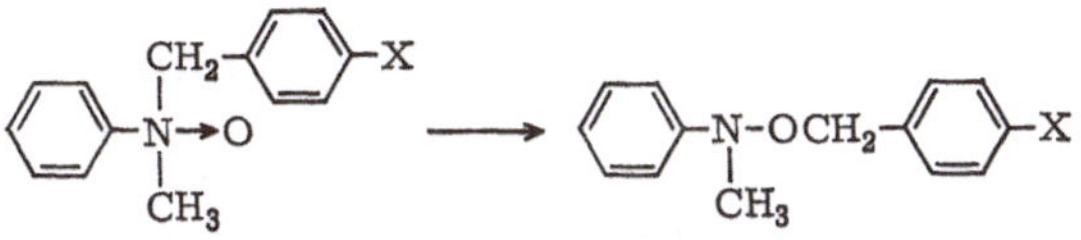

Evidence for the decomposition of a benzyl ylid to an ion pair was reported by *Kantor* and *Hauser* [58] in the reaction of the 2,4,6-trimethyl-benzyltrimethylammonium ion with sodium amide. In this reaction, where the quaternary ion cannot undergo an *ortho*-rearrangement, isodurene was formed in 63% yield along with much formaldehyde.

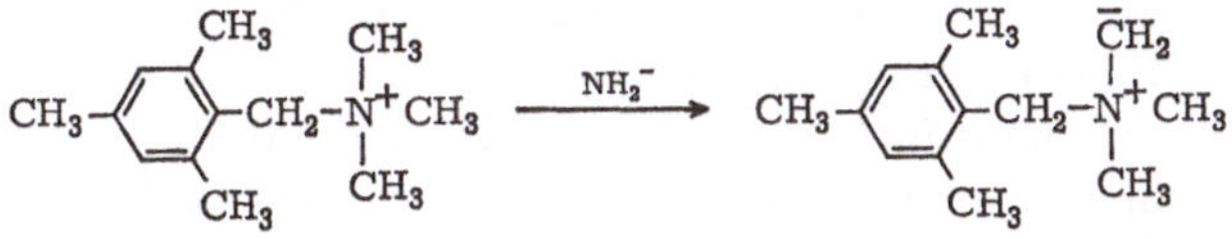

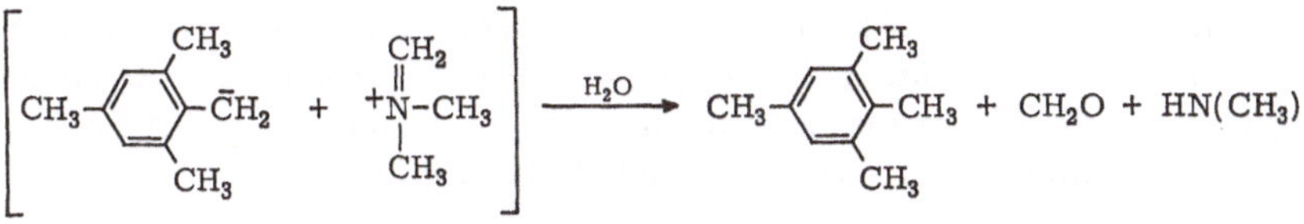

This observation also supports the suggestion that the *exo*-methylene cyclohexadiene is involved in the rearrangement of the benzyltrimethyl-ammonium ion.

Bumgardner [22] also suggested an immonium ion-pair as a possible intermediate in the rearrangement of allylbenzyldimethylammonium-bromide with sodium amide in liquid ammonia. However, he notes that an ylid intermediate must be formed in the rate-determining step and the ylid rearranges directly or gives the intermediate ion-pair which rearranges.

Jenny and *Druey* [77] presented a rather convincing argument for the existence of an ylid breaking down to an ion-pair intermediate in the decomposition of optically active allyl dimethyl(1-phenylethyl)ammo-nium bromide, *24*, with sodium amide in benzene at 80 °C or in liquid ammonia at −33 °C. They observed that the migrating group retains its configuration regardless of whether it migrates to the 2 *(25)* or to the 4

341

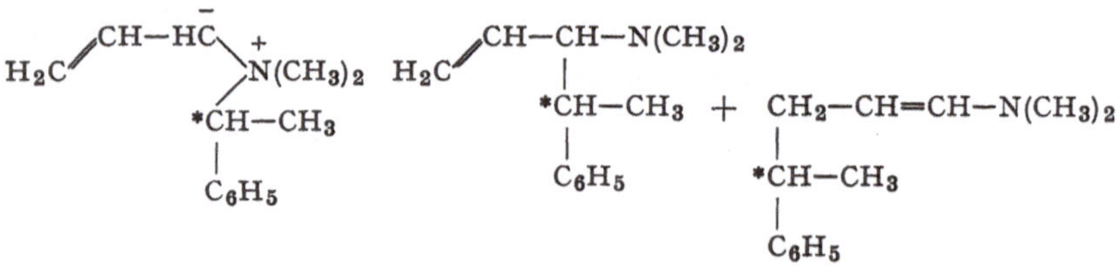

24 25 26

(26) position. They concluded that if only one mechanism leads to the two products, it cannot be an S_Ni mechanism, since this would cause the 1,4 rearrangement product, *26*, to be inverted. Since the configuration of *26* is retained, the benzylic group must migrate to the 4-position with its pair of electrons and without inversion. Similar studies on allyl derivatives have been carried out by *Hellmann* [65] and by *Saunders* and *Gold* [125].

Banthorpe [10] has questioned the conclusions regarding ion pair intermediates in this reaction. He suggests a concerted 1,4 shift to explain the observations of *Jenny* and *Druey* [77] since the C_1—N bond has no stabilization in the transition state. Models also indicate that a 1,4 shift with retention of configuration is not as unfavorable as claimed. According to *Banthorpe* [10] the S_Ni mechanism also seems to account satisfactorily for the reaction of *27* with t-butoxide in dimethylsulfoxide [67]. In this reaction, a compound with an optically active nitrogen is converted to a compound with an optically active carbon. An extremely tight ion-pair intermediate would have to be formed in order for this reaction to proceed with the observed stereospecificity.

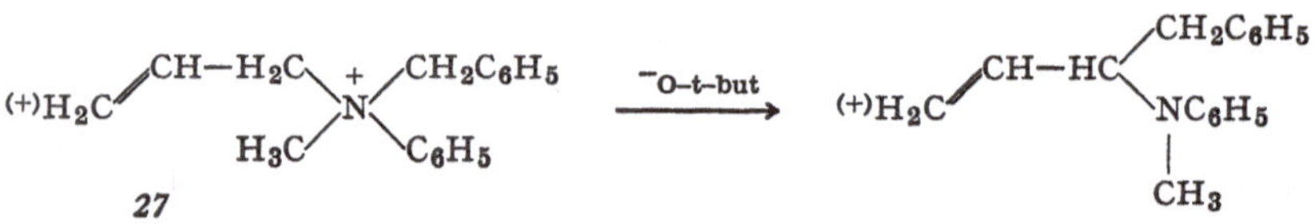

27

An ion pair mechanism has also been invoked to account for the 1,2 and 1,4 rearrangements which are observed when tetrahydropyridinium salts are heated with phenyllithium in ether [76].

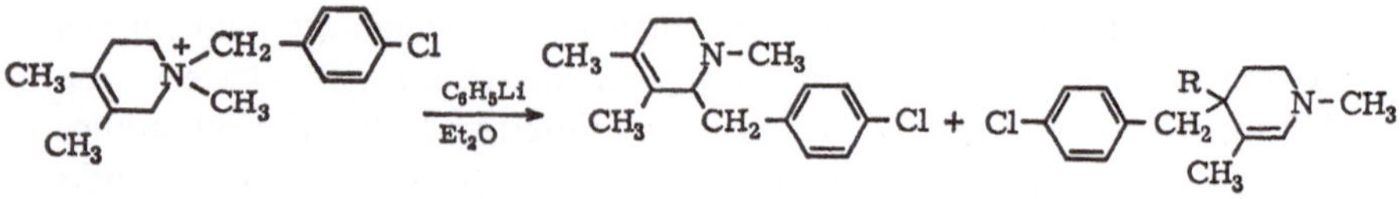

In this case an S_Ni reaction at the 4-position is unlikely for steric reasons.

E. Benzhydryl Substitution

Although the *Sommelet* rearrangement [138] product *29* was initially reported in the decomposition of benzhydryltrimethylammonium hydroxide, *28*, the major products result from displacement reactions [71]. *Hughes* and *Ingold* [71] studied the decomposition of benzhydryltrimethylammonium hydroxide as a function of concentration.

Alcohol Distribution in the Decomposition of Benzhydryltrimethylammonium Hydroxide [71]

Concn. salt	Alc. Total Yield	% ϕ_2CHOH	% MeOH
2.94 N	73	52	49
2.4	93	62	38
1.2	100	85	15
0.7	100	89	11
0.56	100	91	9

In this reaction the formation of methanol was second order, whereas the formation of diphenylcarbinol was first order. This suggests that the ionization of the benzhydryl proton to give an ylid is the rate determining step in the cleavage of this molecule. The ylid must then decompose in a fast step to give product [71].

When *28* is treated with n-butyllithium in ether-hexane [83], with phenyllithium in ether, or with 1:5 phenyllithium: phenylsodium in ether [148], a variety of rearrangement products are formed.

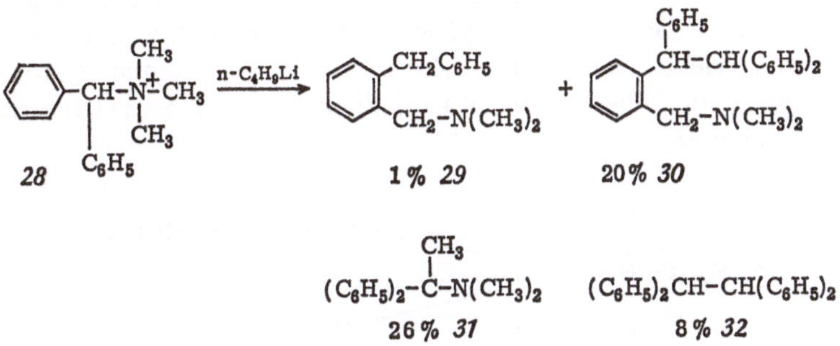

In the reaction of *28* with butyllithium or phenyllithium, the major product, *31*, is reported to result from a *Stevens* rearrangement of the most stable ylid or its lithium derivative [148].

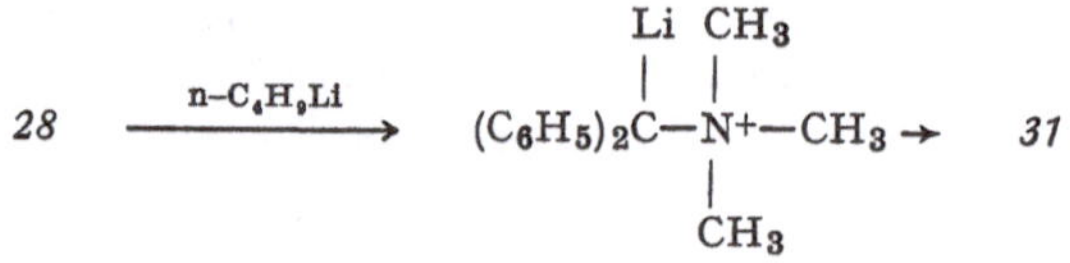

28
n–C₄H₉Li
Li CH₃
(C₆H₅)₂C—N+—CH₃
CH₃
31

The high yield of *30* is ascribed to the displacement of the benzhydryl group by the lithio derivative of *29*. The tetraphenylethane, *32*, is reported to result from a displacement on *28* by benzhydryllithium which may have resulted from the decomposition of the methyl ylid to an immonium ion-pair.

CH₂Li
CH-N±CH₃
CH₃
C₆H₅
CH
CH₂
N±CH₃
CH₃
Li
CH

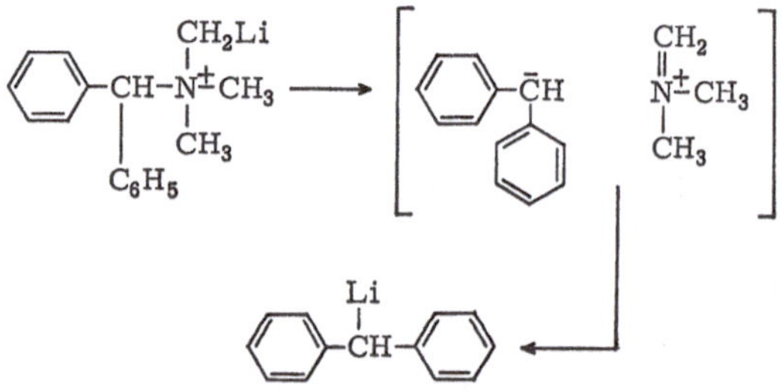

If phenylsodium is used in the reaction along with phenyllithium, the metallated ylid would be less stable, and the products would be those expected from the reaction or decomposition of the free ylid and not of its lithio derivative. When this reaction was carried out by *Tschesche* [148] in the presence of an excess of cyclohexene, the benzhydryl ylid was rapidly formed as indicated by the red color of the solution. However, the color was quickly discharged to give 28% *32*, 27% *31*, 12% *29*, 32% *30*. To account for these results, it was assumed [148] that the benzhydryl ylid is initially formed in 70% yield whereas the methyl ylid is formed in a 30% yield. Half of the benzhydryl ylid decomposes to give diphenylcarbene whereas the other half undergoes a *Stevens* rearrangement to give *31*.

CH₃
(C₆H₅)₂C—N+—CH₃
CH₃
51%
[(C₆H₅)₂C:]+N(CH₃)₃
49%
31

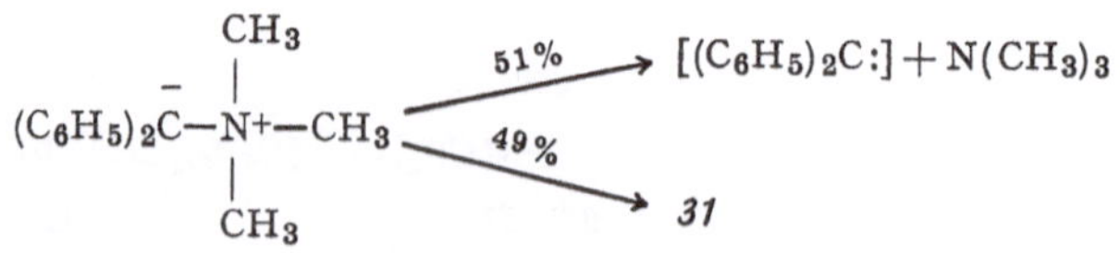

The stabilization of the benzhydryl ylid is expected to be quite high since the related fluorenylid (2) can be isolated as a salt free solid. The high yield of 32 was attributed to hydrogen atom abstraction from the cyclohexene by the diphenylcarbene and subsequent coupling of diphenylmethyl radicals [46].

It is interesting to note that the fluorenylid decomposes to difluorenylidene and trimethylamine on heating [161]. These products may result from the ylid decomposing to the carbene which couples rather than abstracts a hydrogen atom as in the case of benzhydryl carbene in the presence of cyclohexene.

Klein, Van Eenam, and *Hauser* [83] have generalized the reaction of butyllithium with 6, 14, and 28. The benzyl quaternary ion 6 mainly undergoes an *ortho*-rearrangement, the dibenzyl quaternary ion 14 a displacement reaction, and the benzhydryl quaternary ion 28, a *Stevens* rearrangement. The acidity of the benzhydryl proton is responsible for the *Stevens* rearrangement in 28. The reason for the displacement reaction in 14 is attributed to the fact that benzyldimethylamine is a better leaving group than trimethylamine. It is known [3,23] that a phenyl substituted amine is a good leaving group under *Hofmann* conditions; however, the benzyl group has not been studied in these reactions.

Although each of these ions reacts differently toward butyllithium, treatment with phenyllithium always gives predominantly the *Stevens* rearrangement, and treatment with amide ion in liquid ammonia gives predominantly the *ortho*-rearrangement. Sulfonium and benzyl sulfides [59] on treatment with potassium amide undergo a similar *ortho*-rearrangement which proceeds *via* the sulfur ylid.

VII. Ylids from Tetraalkylammonium Salts

The existence of ylids in the decomposition of tetraalkylammonium salts has received some attention in recent years, but only in reactions with organometallic reagents has their presence been demonstrated [6,32,156].

The basic decomposition of tetraalkylammonium salts (the *Hofmann* degradation), has been reviewed extensively [30,135] and will not be discussed here in detail. However, it should be noted that both displacement reactions and α-proton abstraction reactions may occur in addition to elimination reaction [30]. *Ingold* and *Patel* [75] report that the amount of substitution relative to elimination varies depending upon both the substituent on nitrogen and the base.

In all cases the other products are the alcohol and the tertiary amine. It was not determined whether any ethers were produced in these reactions. Likewise proton scrambling experiments have not been carried out

Decomposition of Quaternary Alkylammonium Alkoxides
(% Elimination)

Quaternary ion	OH⁻	OCH₃⁻	OC₂H₅⁻
$\overset{+}{C_2H_5-N-(CH_3)_3}$	94	90	88
$\overset{+}{n-C_4H_9-N(CH_3)_3}$	77	—	—
$\overset{+}{iso-C_4H_9N(CH_3)_3}$	63	57	55

on these molecules and it is not known whether any proton exchange or ylid formation occurs.

The decomposition of decyltrimethylammonium hydroxide at 200 °C as a function of salt concentration was studied by *Hughes* and *Ingold* [72]. The formation of the displacement product, N,N-dimethyldecylamine, is favored in the presence of solvent whereas the elimination product, decene, is favored when the dry salt is decomposed. Thus, β-proton abstraction processes are more favorable as the solvent is removed.

Another interesting comparison was reported by von *Braun* [150] in which various quaternary ammonium hydroxides were decomposed under *Hofmann* conditions and in the presence of glycerol. These results also show that the displacement reaction is favored in the presence of a hydroxylic solvent. It was also noted by *Hanhart* and *Ingold* [57] that displacement reactions are favored over elimination reactions in thermal decompositions of quaternary ammonium salts as the nucleophilicity of the anion is increased from chloride to acetate.

Although the normal course of the *Hofmann* elimination is *via* a concerted *trans* E-2 elimination, in certain cases a *cis* elimination is observed. For example, both *cis*- and *trans*-2-phenylcyclohexyltrimethyl-ammonium hydroxides give 1-phenylcyclohexene on decomposition [4].

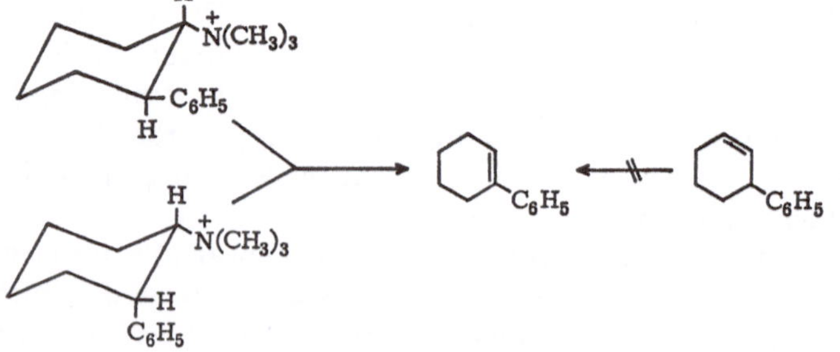

That 3-phenylcyclohexene had formed initially in a *trans* elimination and then isomerized to the product was ruled out, since this compound is stable under the conditions of the reaction [152]. In addition, *trans*-3,3,6,6-tetra-deutero-2-phenyl-cyclohexyltrimethylammonium hydroxide, *33*, decomposed to the olefin *34* without loss of deuterium [29]. A *cis*-elimination involving a methylene ylid could account for the observed reaction.

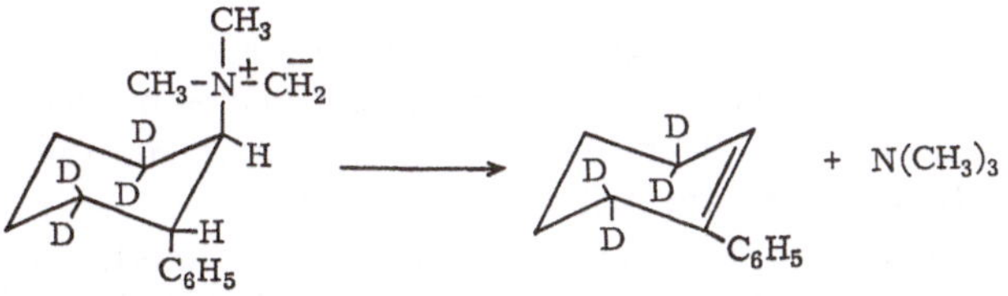

33   *34*

This type of reaction is termed an α′-β elimination and it is isoelectronically similar to the pyrolysis of amine oxides [30]. However, the occurrence of an α′-β elimination was subsequently disproven in this particular reaction because the trimethylamine which forms in the decomposition of *trans*-2-phenyl-2-deuterocyclohexyltrimethylammonium hydroxide contains no deuterium [7]. Therefore, a *cis* E-2 mechanism was proposed to account for the experimental observations [36].

The α′-β elimination was initially proposed by *Wittig* and *Polster* [171] to account for the formation of the trimethylamine, propylene and iodobenzene when isopropyldimethyliodomethylammonium iodide was treated with phenyllithium.

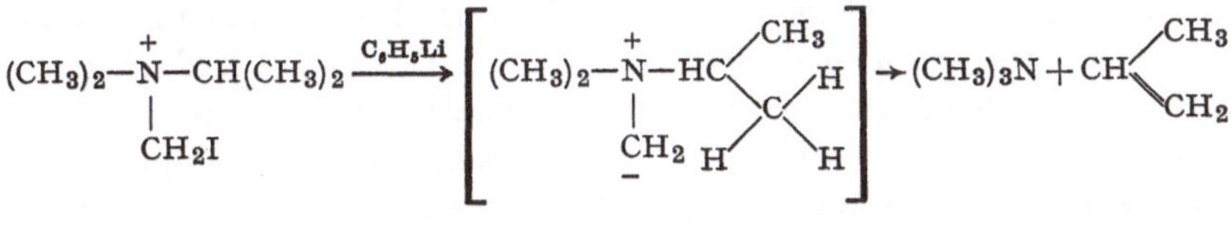

To determine whether the α′-β mechanism was operating in the *Hofmann* degradation, several labelling experiments were carried out by *Weygand, Daniel*, and *Simon* [156]. They examined the decomposition of 2-tritioethyltrimethyl ammoniumhydroxide at 150 °C and the reaction of the bromide salt with phenyllithium at room temperature. Since only a small amount of tritium was found in the trimethylamine when the hydroxide was decomposed, they concluded that the α′-β path was not important. However, 54% of the original activity was present in the trimethylamine when the bromide salt was treated with phenyllithium, suggesting that the α'-β path with an ylid intermediate was an important decomposition route in the presence of an organometallic reagent [156]. Results of the pyrolysis of β-d₃-ethyltrimethylammonium hydroxide also indicated that an ylid path was not important [32].

In a similar reaction, e.g., the decomposition of cyclohexylmethyl-β-d-trimethylammonium salts, analogous results were observed. The α'-β path was important when phenyllithium was used, but this route could not be detected in the *Hofmann* degradation of the hydroxide [32]. However in the *Hofmann* degradation at 95 °C, only 45% of the reaction proceeds by elimination. The other 55% decomposes by demethylation. The elimination does not proceed by a α'-β process, since no deuterium was initially found in the trimethylamine. However, as the reaction proceeds, up to 20% deuterium is found in the trimethylamine. Thus, an exchange reaction between the solvent (H₂O and DOH formed during the abstraction reaction) and the methyl carbon atoms of the quaternary ammonium salt occurs simultaneously with the elimination and displacement reactions. This exchange reaction was postulated to account for the incorporation of deuterium into the trimethylamine as the reaction proceeds rather than the initiation of an α'-β elimination. Confirmation of the incorporation of deuterium in the trimethylamine by exchange with solvent in the quaternary ammonium salt was obtained by comparing the amount of deuterium in the N,N-dimethyl-cyclohexylmethyl-β-d amine formed in the displacement reaction with the amount of deuterium in the trimethylamine formed in the elimination reaction as the reaction proceeds [32]. Since both of these products must come from the same starting material, with a statistical distribution of deuterium in the methyl groups, this technique constitutes a reliable check on an exchange mechanism occurring simultaneously with elimination. Such an exchange reaction proceeding along with elimination has been invoked a number of times to rule out an α'-β elimination in the *Hofmann* degradation even though some deuterium is found in the trimethylamine. It must be emphasized, however, that this conclusion is based on the assumption that the mechanism does not change as the "solvent" is removed during the course of reaction. If the exchange reaction proceeds *via* an ylid intermediate, reprotonation by the solvent will be reduced as the solvent is removed, and abstraction of the β-hydrogen may become favorable. Although this possibility always exists, an α'-β elimination has never been definitely characterized in hydroxylic solvents.

Recently it was reported [26,28] that the existence of a *cis*- or *syn*-elimination can *never* be used to show that an α'-β mechanism is operating since a *cis* E-2 elimination with concerted but not a synchronous bond breaking is also possible and appears to be more favorable than the α'-β path. Decomposition of N,N,N-trimethyl-3-*exo*-d₁-bicyclo[2.2.1]heptyl-

2-exo ammonium hydroxide, *35,* gives bicyclo[2.2.1]hept-2-ene, *36,* containing no deuterium [26].

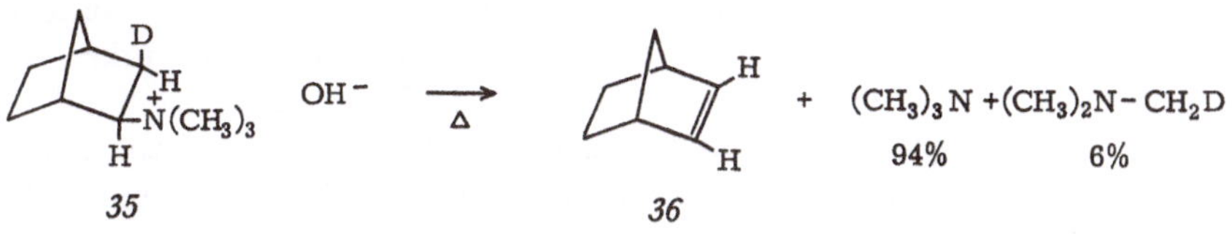

$$\text{(structure 35)} \quad OH^- \xrightarrow{\Delta} \quad \text{(structure 36)} \quad + \quad (CH_3)_3N + (CH_3)_2N-CH_2D$$

$$35 \qquad\qquad\qquad 36 \qquad\qquad 94\% \qquad 6\%$$

Initially the trimethylamine contains 6% deuterium and it was concluded that a maximum of 6% of the reaction proceeds thru an ylid path [26]. However, when the reaction is carried to completion the trimethylamine contains 17% trimethylamine-d_1. In this reaction no substitution products (dimethylalkylamines) were obtained when the hydroxide was dried thoroughly, under high vacuum, but an exchange reaction which incorporates deuterium into the methyl groups must be operating and a concerted exchange would have to involve at least one mole of solvent. However, if ylid formation is proposed to account for the exchange reaction, an α'-β elimination reaction must not occur.

When steric factors prevent the attack of base at the β-hydrogen as shown the following reaction, the *Hofmann* degradation is found to proceed *via* an α'-β path [31].

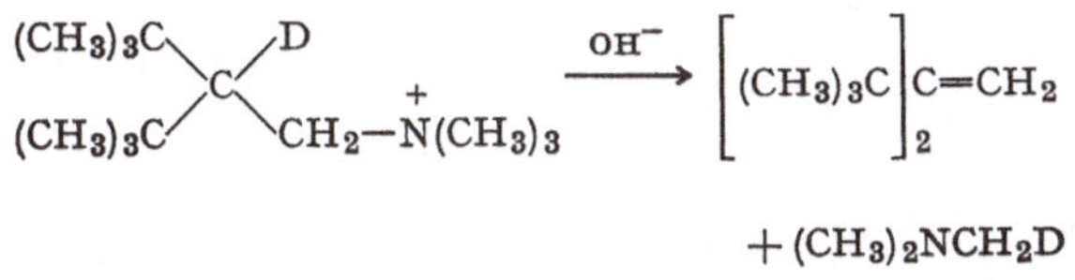

$$\begin{array}{c} (CH_3)_3C \\ C \\ (CH_3)_3C \end{array} \hspace{-0.5em} \begin{array}{c} D \\ \\ CH_2-N(CH_3)_3 \end{array} \xrightarrow{OH^-} \left[(CH_3)_3C \middle| C=CH_2 \right]_2$$

$$+ (CH_3)_2NCH_2D$$

VIII. Ylids from Cyclic Quaternary Ammonium Salts

A series of bases was used to cause the elimination of trimethylamine from the cyclooctyltrimethylammonium ion [174]. The thermal decomposition of the hydroxide at 140 °C or reaction with sodium amide at −40 °C gives a mixture of *cis-* and *trans-*cyclooctenes with the *trans-*form predominating and suggesting a normal *trans* E-2 elimination. However, when phenyllithium is used as the base, the *cis-*cyclooctene predominates. The *cis-*cyclooctene also occurs predominantly when bromomethyldimethyl cyclooctylammonium bromide is treated with phenyllithium. Thus, the ylid path operates when organolithium reagents are used in nonpolar solvents and the normal *trans-*E-2 elimination occurs in hydroxylic solvents. If the α-hydrogens are made more acidic, as in the benzyldimethylcyclooctylammonium ion, treatment with sodium amide in liquid ammonia gives *cis*-cyclooctene exclusively [21]. Thus the conversion from a β-elimination to an α'-β elimination by a subtle change in the structure of the cation was apparently achieved [21]. However, as discussed in Sect. 7, the presence of *cis*-cyclooctene cannot be taken as conclusive evidence that an α'-β elimination had occurred.

In other cyclic systems, it is found that the reaction of the N,N-dimethylpyrrolidinium cation with hydroxide or amide gives 4-dimethylaminobutene-1 in a normal *Hofmann* elimination [174] However, a fragmentation reaction giving ethylene and dimethylvinyl amine occurs

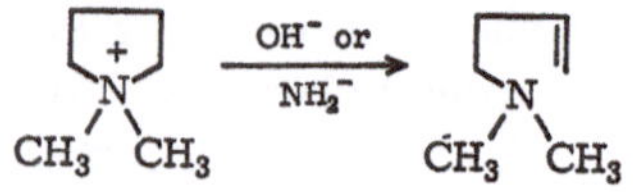

37

during the reaction of the dimethylpyrrolidinium ion with butyllithium or phenyllithium [38,154] in ether. The dimethylvinylamine is not isolated, but decomposes to dimethylamine and acetaldehyde on hydrolysis.

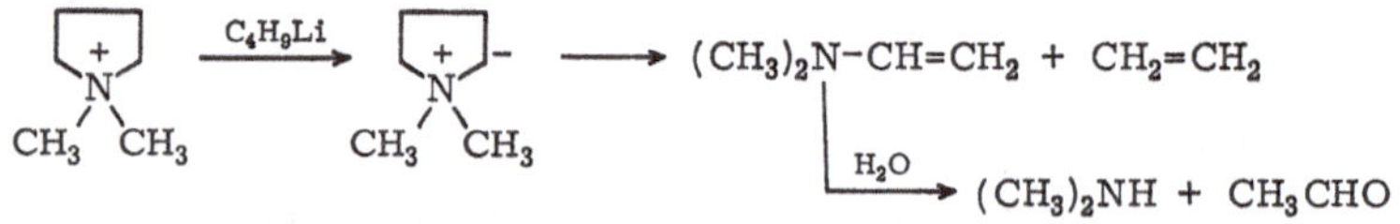

37

The reaction probably proceeds thru the ylid generated in the ring, since the methylene ylid, prepared from N-methyl-N-bromomethyl-pyrrolidinium bromide and butyllithium, does not undergo the α' β elimi-

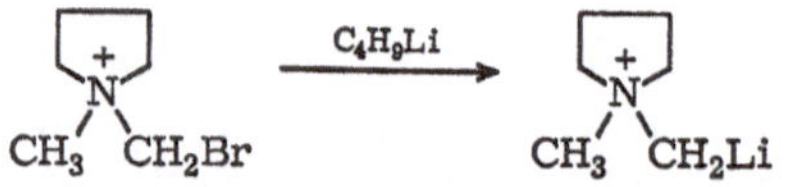

nation to give 4-dimethylaminobutene-1. The methylene ylid therefore decomposes to N-methylpyrrolidine and dimethylamine [175]. The dimethylamine must result from a rearrangement of the methylene ylid to the ring ylid which decomposes by the fragmentation reaction [175]. When N,N-dimethyl-2-phenylpyrrolidine is treated with phenyllithium, products resulting from substitution and *ortho*-rearrangement paths are observed, as well as fragmentation [37].

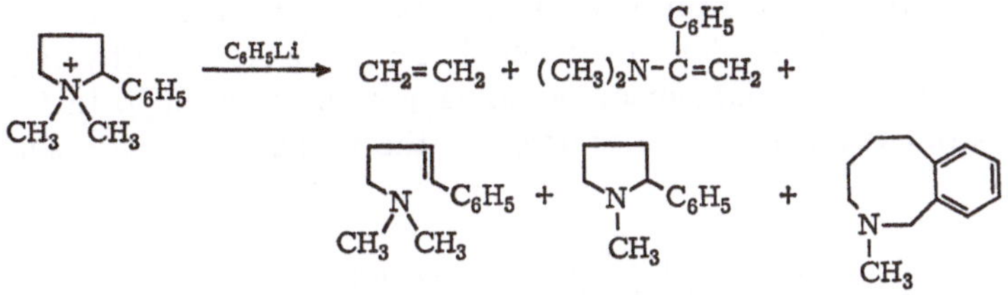

The decomposition of N-trideuteromethylquinuclidinium methoxide *(38)* was studied in an attempt to induce a simple demethylation without hydrogen scrambling. However, when the pyrolysis was carried out, the major product (92%) was the ring opened piperidine *(39)*, which appears

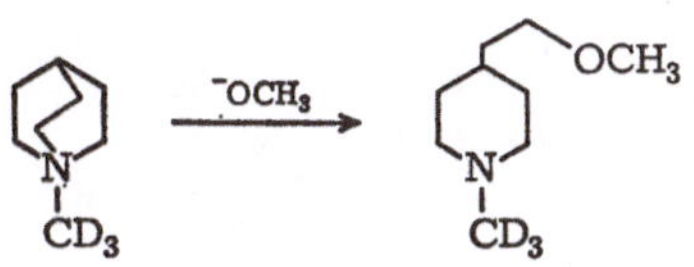

to be formed in a simple displacement reaction [111]. In the decomposition of N-methyl quinuclidinium hydroxide, three products are formed in low yield: quinuclidine, 4-vinyl-N-methyl piperidine, and 4-(2-hydroxy-ethyl)-N-methyl piperidine [102].

Anderson and *Wills* [1] report ylid intermediates in the reaction of azetidines with potassium amide in liquid ammonia. The major product resulting from the reaction with 1,1,3,3-tetramethylazetidinium bromide, *40*, is the ring enlarged product *41* in 70% yield. Both an ion pair mech-

anism and a carbene mechanism were suggested to account for the high yield of the pyrrolidine, but the ion-pair mechanism was favored.

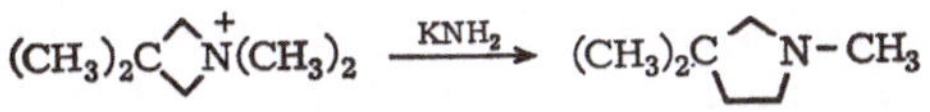

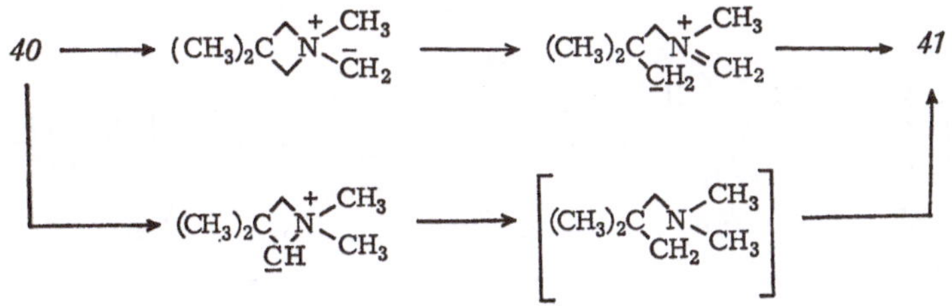

Alternative concerted processes do not appear to release much ring strain in the activated complex, and it would be expected that methyl migration rather than ring enlargement would have been observed [1]. In 1-benzyl-1,3,3-trimethylazetidinium iodide, *42*, the major product is again the

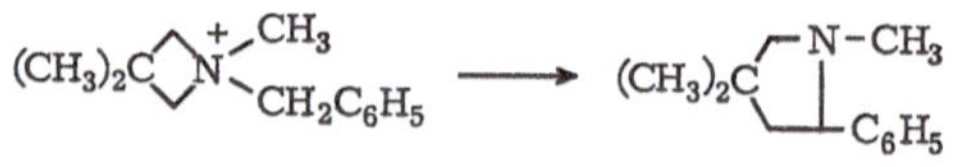

42　　　　　*43*

pyrrolidine. However, in the N,N-dibenzyl derivative, *44*, the major product is the *Sommelet* product, *45*, in 98% yield with only a small amount

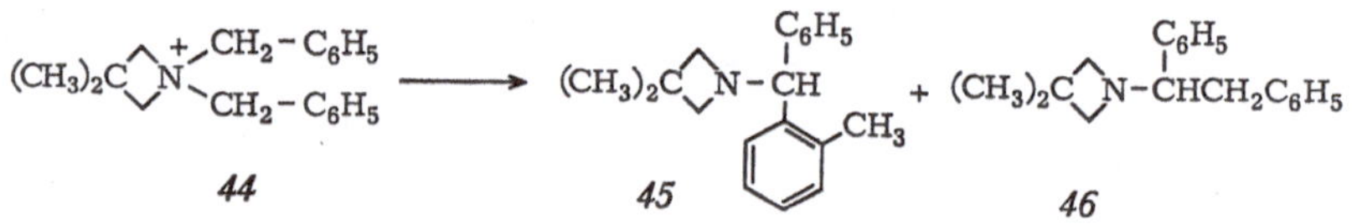

44　　　　　*45*　　　　　*46*

(2%) of the *Stevens* rearrangement product, *46*. Apparently the ion-pair involving the benzyl carbanion is more stable than the ring opened carbanion and the reaction proceeds to give *45* rather than the pyrrolidine.

IX. Other Synthetic Approaches to Nitrogen Ylids

Although the most common method of preparing ylids involves proton abstraction from a quaternary ammonium salt, another method which has been used in ylid synthesis involves the reaction of carbenes with tertiary amines. *Bamford* and *Stevens* [8] reported that the reaction of diazofluorene, *47*, with benzyldimethylamine at 150 °C in the absence of solvent gives 9-benzyl-9-dimethylaminofluorene in 30% yield *(48)*. Later, *Wittig* and *Schlosser* [177] noted that if copper II was used as a catalyst, the reaction could be carried out at a lower temperature in THF to give a 21% yield of the same product.

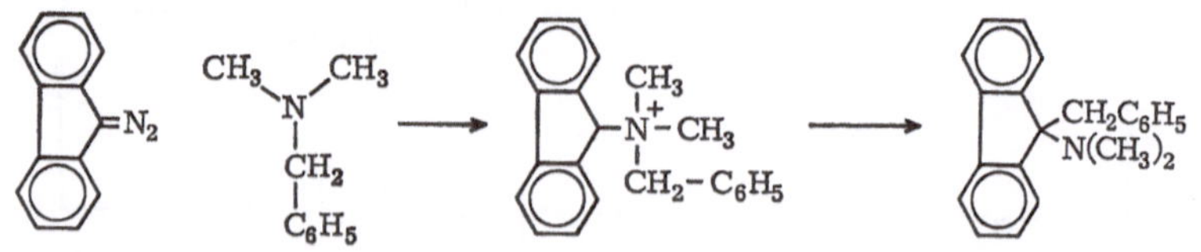

47　　　　　*48*

It was also shown [48] that when trimethylammonium fluorenylid was treated with dimethylbenzylamine, trimethylamine was eliminated and *48* was obtained. A transamination involving an intermediate carbene was suggested as the precursor to *48* [48].

Saunders and *Murray* [124] found that if tertiary amines are treated with dichlorocarbene (generated from potassium *t*-butylate and chloroform in benzene) an ylid intermediate is formed. For example, when benzyldimethylamine is treated with dichlorocarbene, N,N-dimethylphenylacetamide and dibenzyl are isolated. The following mechanism was proposed for the formation of the amide.

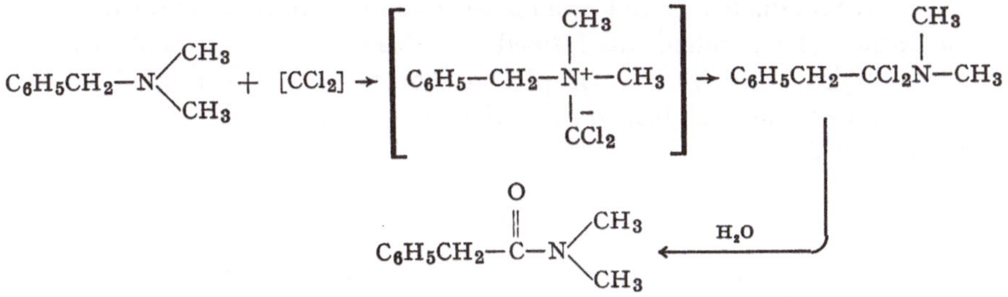

$$C_6H_5CH_2-N\begin{smallmatrix}CH_3\\\\CH_3\end{smallmatrix} + [CCl_2] \rightarrow \left[C_6H_5-CH_2-\overset{\overset{CH_3}{|}}{\underset{\underset{CCl_2}{|^-}}{N^+}}-CH_3 \right] \rightarrow C_6H_5CH_2-CCl_2\overset{\overset{CH_3}{|}}{N}-CH_3$$

$$C_6H_5CH_2-\overset{\overset{O}{||}}{C}-N\begin{smallmatrix}CH_3\\\\CH_3\end{smallmatrix} \xleftarrow{H_2O}$$

No ortho rearrangement product was observed. The presence of dibenzyl in this reaction is surprising. It was proposed that the ylid decomposed to an ion-pair containing a benzyl carbanion which is then alkylated by the protonated ylid to give dibenzyl [124].

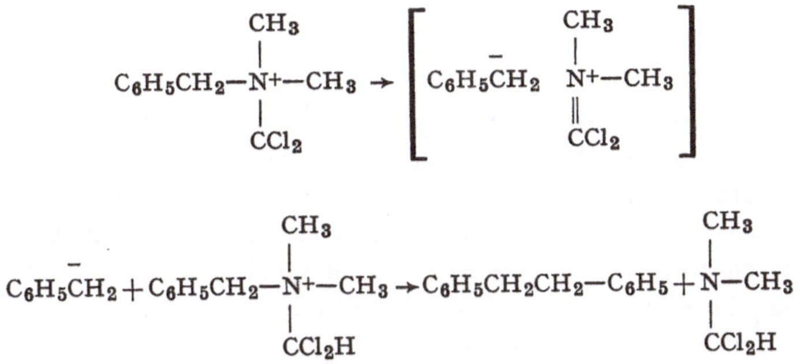

$$C_6H_5CH_2-\overset{\overset{CH_3}{|}}{\underset{\underset{CCl_2}{|}}{N^+}}-CH_3 \rightarrow \left[C_6H_5\overset{-}{C}H_2 \quad \overset{\overset{CH_3}{|}}{\underset{\underset{CCl_2}{||}}{N^+}}-CH_3 \right]$$

$$C_6H_5\overset{-}{C}H_2 + C_6H_5CH_2-\overset{\overset{CH_3}{|}}{\underset{\underset{CCl_2H}{|}}{N^+}}-CH_3 \rightarrow C_6H_5CH_2CH_2-C_6H_5 + \overset{\overset{CH_3}{|}}{\underset{\underset{CCl_2H}{|}}{N}}-CH_3$$

The *t*-butanol was suggested as the source of the hydrogen required to protonate the ylid [124]. However, in the presence of *t*-butanol, the free benzyl carbanion should have also been protonated to give toluene, but no toluene was observed. Perhaps a radical path involving the coupling of benzyl radicals would account for dibenzyl in the most straightforward manner.

The reaction of dichlorocarbene with triethylamine resulted in α'-β elimination of ethylene and the formation of diethylformamide after hydrolysis [124]. When trimethylamine was treated with dichlorocarbene, no normal *Stevens* rearrangement products were observed and the major product isolated was α,α,β-trichloroethylamine. The presence of this extra chlorine atom in the product is unexpected and a satisfactory mechanism has not been proposed for its formation.

The reaction of carbene (from diazomethane) with N-methyl pyrrolidine gives products resulting from insertion into all of the C-H bonds but no N-methyl piperidine was formed by a *Stevens* rearrangement of the methyl ylid. However, when phenyl diazoacetate was treated with benzyldimethylamine, benzyl migration occurs presumably *via* an intermediate ylid [45].

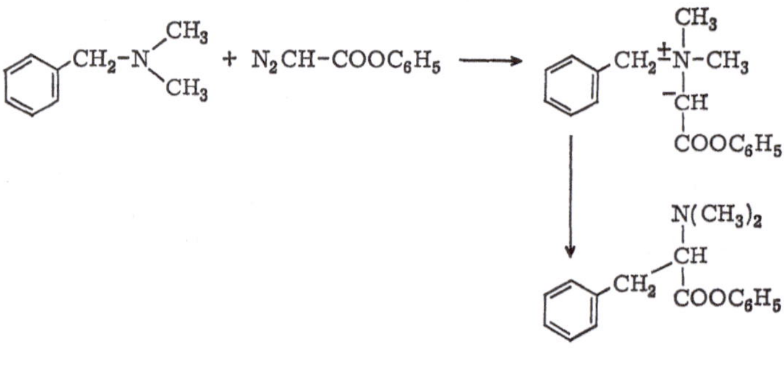

Another method of generating ylids developed by *Wittig* and coworkers [160,173] and by *Hellmann* and *Unseld* [64] involves the reactions of tertiary amines with benzynes. Recently, *Lepley, Becker,* and *Giumanini* [93] studied the reaction of benzyne with N,N-dimethylbenzylamine and showed that the major product of the reaction was N-methyl-N-(α-phenethyl)aniline, in 35% yield. Small amounts of N-methyl-N-benzyl-aniline and N-methyl(β-phenethyl)aniline were also observed. To explain these observations it was suggested that the phenyl carbanion which is formed initially undergoes an internal proton migration to give the benzyl or methylene ylid. Subsequent rearrangement of the ylids gives rise to the observed products.

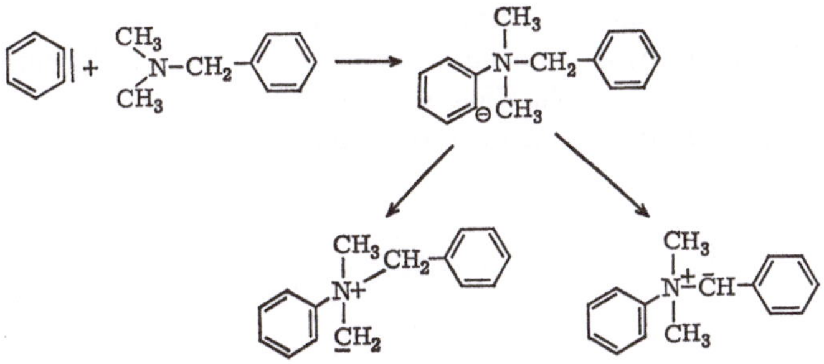

Proton magnetic resonance studies of the reaction showed stimulated emission and enhanced absorption of the benzylic protons on N-methyl-N-(α-phenylethyl)-aniline. This observation suggests that the benzyl ylid untergoes a *Stevens* rearrangement by a free radical pair methyl migration [94,99].

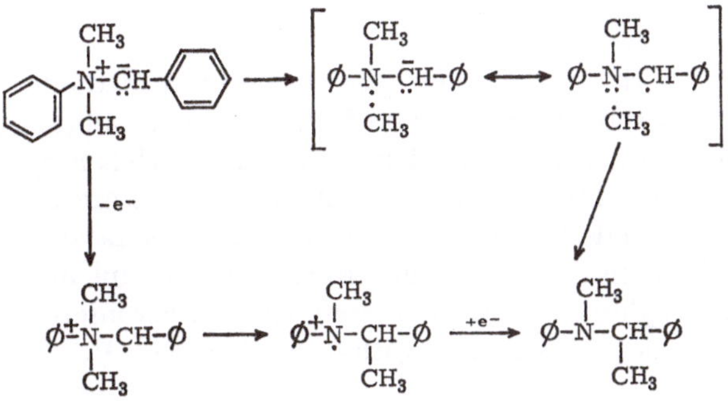

This radical-pair mechanism may be extremely important in many ylid rearrangement processes.

X. Other Nitrogen Ylids

In the preceeding discussion those ylids which are stabilized by delocalization of the cationic and/or anionic charge were not treated extensively. For completeness, a few of these compounds will be described here.

The synthesis and reactions of trimethylammonium fluorenylid 2, an isolable solid, has already been discussed. Trimethylammonium cyclopentadienylid has also been prepared [41] as a pink solid, but no reactions of this compound have been reported. However, it might be expected that its behavior would be similar to that of the fluorenylid.

One of the most stable ylids is *trimethylammonium dicyanomethylid*. This ylid is stable at room temperature and in the presence of oxygen and

$$(CH_3)_3\overset{+}{N}-\overset{-}{C}\diagup^{CN}_{\diagdown CN}$$

water, and melts at 154 °C. Although this ylid appears similar to the dichloromethylid described earlier, no corresponding reactions of this ylid have been reported.

Pyridinium ylids can be stabilized if the anionic portion of the ylid is delocalized [42,77].

$$\overset{\diagup\diagdown}{\underset{\diagdown\diagup}{}}N^{\oplus}-\overset{-}{C}H-R \qquad R = -COC_6H_5, \ CSSCH_3, \ etc.$$

Much of the work on pyridinium ylids has been carried out by *Kröhnke* and his papers should be consulted for background information on this subject [90,91]. A general discussion of pyridinium ylids is also presented by *Johnson* [77]. Alkylation and acylation reactions are quite common with this ylid and it has been shown that C-alkylation followed by reductive cleavage of the pyridine residue is a useful method for the synthesis of alkylated ketones and esters [66]. Another interesting elimination occurs if the ylid is treated with nitrosobenzene [89].

$$\overset{\diagup\diagdown}{\underset{\diagdown\diagup}{}}N^{\pm}\overset{-}{C}H-\overset{O}{\overset{\|}{C}}-C_6H_5 + C_6H_5NO \longrightarrow \overset{\diagup\diagdown}{\underset{\diagdown\diagup}{}}N + C_6H_5N=CH-\overset{O}{\overset{\|}{C}}-C_6H_5$$

This type of reaction has not been reported for *1*.

A rather interesting synthesis of a pyridinium ylid was reported by *Phillips* and *Ratts* [113]. The reaction of pyridine with bromoacetic acid in the presence of benzaldehyde at 120 °C in nitrobenzene solvent gives 1-(2-hydroxy-2-phenylethyl)pyridinium bromide in 75% yield. The mechanism suggested for the reaction involves the formation of the the pyridinium betaine which decarboxylates to the ylid. Subsequent reaction of the ylid with benzaldehyde and protonation gives the final product.

$$2 \text{ (pyridine)N} + BrCH_2COOH \longrightarrow \text{(pyridinium)}\overset{+}{N}\text{-}CH_2\text{-}COO^- + \text{(pyridinium)}\overset{+}{N}H + Br^-$$

$$\downarrow -CO_2$$

$$\text{(pyridinium)}\overset{+}{N}\text{-}CH_2\text{-}\underset{\overset{|}{H}}{\overset{\overset{|}{OH}}{C}}\text{-}C_6H_5 \underset{H^+}{\overset{\phi CHO}{\longleftarrow}} \text{(pyridinium)}\overset{+}{N}\text{-}\overset{-}{C}H_2 \longleftarrow$$

It has been suggested [182] that betaine itself may lose CO to give the ylid, but no evidence has been reported to substantiate this idea [181, 183]. Perhaps the most stable ylids are pyridinium cyclopentadienylid [86, 121] and pyridinium dicyanomethylid [100].

$$\text{(pyridinium)}\overset{\oplus}{N}\text{-}\overset{\ominus}{\text{(cyclopentadienyl)}} \qquad \text{(pyridinium)}\overset{+}{N}\text{-}\overset{-}{C}\overset{CN}{\underset{CN}{}}$$

The crystal structure of the dicyanomethylid has been determined [19, 20] and the most unexpected result obtained is that the $C\text{-}(CN)_2$ portion is slightly bent out of the plane formed by the rest of the molecule. Although these ylids can be isolated, their chemical reactivity has not been explored extensively.

The main reaction of the dicyanomethylid is as a 1,3-dipolar species [101]. The reaction with diethyl acetylenedicarboxylate is representative of this behavior.

$$\text{(pyridinium)}\overset{+}{N}\text{-}\overset{-}{C}\overset{CN}{\underset{CN}{}} + \underset{C}{\overset{C\overset{COOEt}{}}{\underset{COOEt}{\parallel}}} \longrightarrow \text{(indolizine)} - COOMe$$

XI. Summary and Conclusions

In summarizing the subject of nitrogen ylids, the data collected from various studies of the reaction of strong bases with quaternary ammonium salts will be examined. In this situation where different investigators have proposed ideas on various aspects of the problem, a unified approach to the entire subject should be suggested. The generality of ideas presented here must be tested in the laboratory, but if they spur additional research, and if a few prove useful, these suggestions will be worthwhile.

When quaternary ammonium salts are treated with strongly basic reagents, proton abstraction from a carbon atom adjacent to the nitrogen atom often occurs. Displacement reactions at the α-carbon atom occur concommitant with α-proton abstraction and the extent of this competetive path depends on the nucleophilicity of the reagent, the acidity of the α-hydrogen, the temperature and the solvent.

If *Lewis* acids are present, the ylid may be stabilized and decomposition may be reduced or prevented entirely. The lithium ion is especially useful in stabilizing the ylid although other *Lewis* acids may be employed. Proton exchange reactions between the ylid and the solvent (*Brønsted* acid) compete with alternative decomposition reactions and tend to stabilize the ylid. It is suggested that these proton exchange reactions generally proceed through an ylid intermediate rather than through an activated complex which resembles that of an S_N2 reaction at a neopentyl carbon.

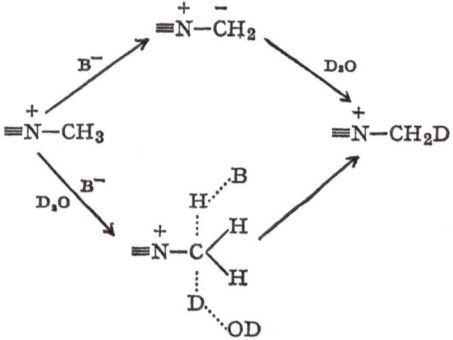

The greater the stabilization of the carbanionic portion of the ylid, *e.g.*, benzyl, the more likely the ylid path is followed.

Once the ylid has been formed, various decomposition reactions are observed. In the presence of *Lewis* acids it is difficult to ascertain whether the free ylid, its metalated adduct, or a dimetalated species is involved in the activated complex leading to decomposition. If the anionic portion of the ylid is stabilized by resonance, the extent of complexation with the *Lewis* acid is reduced.

The most common reactions of the ylid (stabilized or not) are proton abstraction and displacement reactions. In general, proton abstraction reactions are favored, since the ylid can be envisaged as a neopentyl carbanion with a high steric requirement and high basicity.

The following decomposition routes have been observed for nitrogen ylids. These diagrams only illustrate the mode of decomposition and do not imply a description of the species involved in the transition state.

1. α'-β Elimination

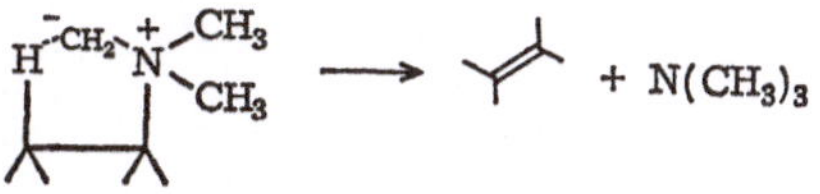

$$H^{-CH_2}\overset{+}{N}\overset{CH_3}{\underset{CH_3}{}} \longrightarrow \bigvee + N(CH_3)_3$$

2. Ion-pair Decomposition

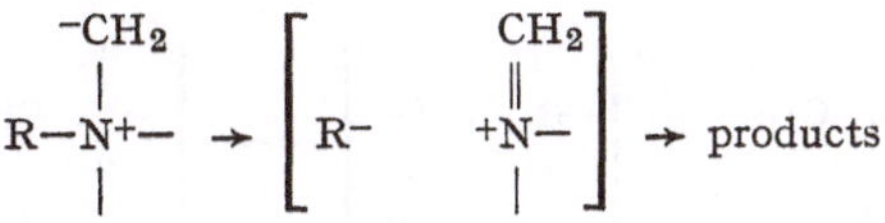

$$R-\overset{-CH_2}{\underset{|}{N^+}}- \rightarrow \left[R^- \quad \overset{CH_2}{\underset{|}{^+N}}- \right] \rightarrow products$$

3. γ-Carbanionic Attack (*Sommelet-Hauser* Rearrangement; *ortho*-rearrangement)

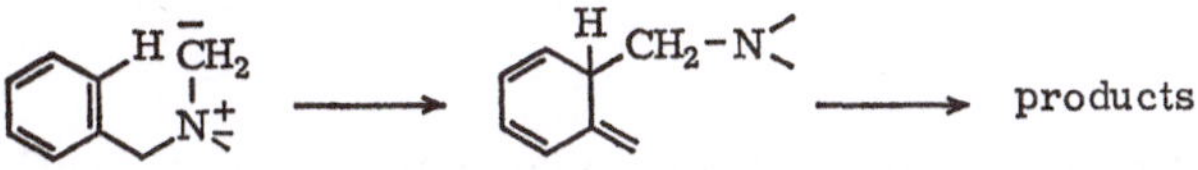

$$\overset{H}{\underset{N^+}{\overset{-}{C}H_2}} \longrightarrow \overset{H}{\overset{CH_2-N}{}} \longrightarrow products$$

4. α-Carbanionic Attack (*Stevens* Rearrangement)

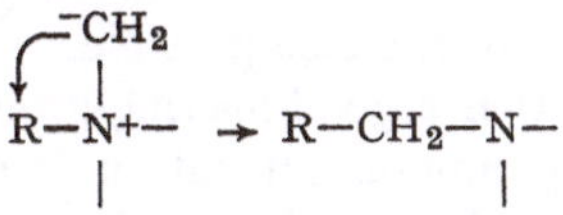

$$R-\overset{-CH_2}{\underset{|}{N^+}}- \rightarrow R-CH_2-\overset{|}{N}-$$

5. α-Elimination (Carbenoid)

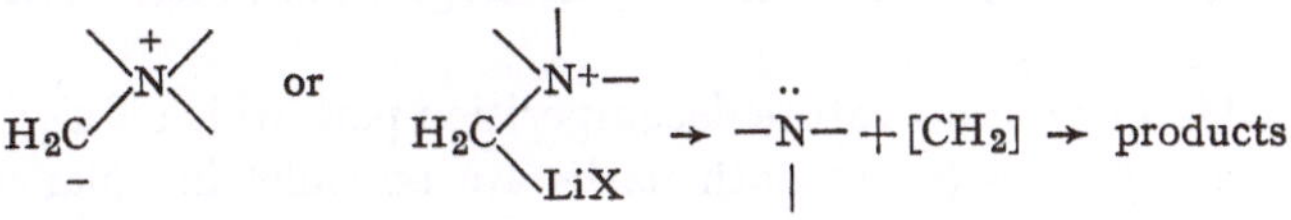

$$H_2C\overset{+}{\underset{-}{N}} \quad or \quad H_2C\overset{N^+-}{\underset{LiX}{}} \rightarrow -\overset{\cdot\cdot}{N}-+[CH_2] \rightarrow products$$

6. Intramolecular Exchange

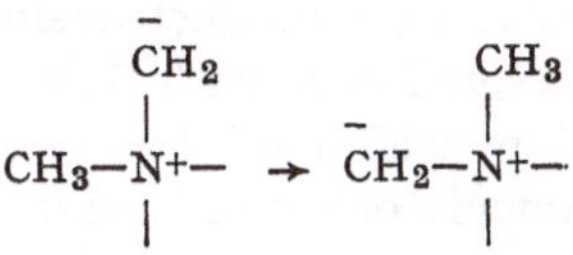

$$CH_3-\overset{\overset{-}{C}H_2}{\underset{|}{N^+}}- \rightarrow \overset{CH_3}{\underset{|}{\overset{-}{C}H_2-N^+}}-$$

7. Intermolecular Exchange

$$
\overset{\scriptstyle -}{\underset{\textstyle |}{\overset{\textstyle |}{CH_2}}}\quad\quad\quad\quad \overset{\textstyle CH_3}{\underset{\textstyle |}{}}
$$

$$
CH_3-\overset{|}{\underset{|}{N^+}}-\ +\ CH_3-\overset{+}{N}\equiv\ \rightarrow\ CH_3-\overset{|}{\underset{|}{N^+}}-\ +\ \overset{-}{CH_2}-\overset{+}{N}\equiv
$$

8. Free Radical Rearrangement

$$
-\overset{|}{\underset{\underset{\textstyle CH_3}{|}}{N^+}}-\overset{-}{CH}-\ \rightarrow\ \left[-\overset{|}{N}-\overset{-}{CH}-\ \atop {\overset{+\cdot}{}\ \ \ \cdot CH_3} \right]\ \longleftrightarrow\ \left[-\overset{|}{N}-\overset{}{CH}-\ \atop {\overset{..}{\downarrow}\ \ \cdot CH_3} \right]
$$

$$
-\overset{|}{\underset{\underset{\textstyle CH_3}{|}}{N}}-\overset{}{CH}-
$$

Path 1 may be important if the ylid cannot exchange with a protonic solvent and a β-hydrogen atom is present. This path has not been definitely observed in the presence of hydroxylic solvents (even in the "dry" state) and is most often observed when organolithium reagents are used as the base. A *syn*-elimination cannot be taken as proof that this path is operating, for *cis* E-2 reactions are also possible.

Path 2 is observed if either a good leaving group (*e.g.*, R⁻ = benzyl) is present or the resulting immonium cation is stabilized. This path should be examined in detail in each reaction since a concerted E-2 mechanism gives rise to the same ion pair without the intermediate existence of a free ylid. The extent of C-H *vs*. N^+-C bond breaking in the transition state must be considered to distinguish between these possibilities.

Path 3 is a low-temperature decomposition path which is significantly favored over path 4 (when both paths are possible) in polar solvents. In non-polar solvents, variation in extent of paths 3 and 4 depend on a variety of factors such as the specific base, the solvent, and the anion. The products resulting from this path may also be formed *via* the rearrangement of an intimate ion-pair intermediate (path 2). It is difficult to separate these extreme cases from product analysis alone.

Path 4 is generally favored over path 3 in aprotic solvents and at elevated temperatures. Here again path 2 may provide the intermediate for the reaction if the anionic or cationic partners of the ion-pair are stabilized under the conditions of the reaction. This path or a more detailed description of α-migration is probably the most common decomposition route of all nitrogen ylids.

Path 5 has not been confirmed in the ylid, but products which can be attributed to this path are noted in the decomposition of the fluorenylid and in alkoxy-substituted quaternary ammonium salts. Although ethylene and polymethylene are observed in the decomposition of the ylid, it has been suggested that these products are formed *via* a stepwise alkylation reaction. However, the ylid may be regarded as a carbenoid, with the $N^+(CH_3)_3$ group behaving as a pseudohalogen, which reacts either as the free ylid or as the lithium halide complex.

Paths 6 and 7 occur concurrently with other decomposition schemes but may remain unnoticed unless deuterium is incorporated into the molecule. It is felt that the intramolecular path is probably the more important route. However, it should be pointed out that a free ylid need not necessarily be present in this case, for a concerted exchange can also occur.

In an nmr study of the reaction of N,N-dimethylbenzylamine with benzyne, *Lepley* [98] observed a free radical intermediate (path 8) which is formed from the ylid and which rearranges to N-methyl-N-(α-phenethyl)aniline *via* a *Stevens* rearrangement. Prior to this report [98,99] radical intermediates had not been observed in the *Stevens* rearrangement but their presence had been suggested [131]. This radical path must be considered in all reactions of nitrogen ylids for it may be the route by which many rearrangement reactions occur.

Acknowledgement: I wish to thank the National Institute of Health for support of much of my research on nitrogen ylids. I wish to thank Dr. Stanley L. Pine for a preprint of a review article entitled, "The Basic Rearrangement of Quaternary Ammonium Salts," which will soon be published in Organic Reactions. I especially thank Dr. Arthur R. Lepley for several preprints of his work on radical intermediates and for his valuable comments on the manuscript.

W. K. Musker

XII. References

1) *Anderson, A. G., Jr.*, and *M. T. Wills:* J. Org. Chem. *33*, 3047 (1968).
2) *Appel, R.*, *H. Heinen* u. *R. Schöllborn:* Chem. Ber. *99*, 3118 (1966).
3) *Archer, D. A.*, and *H. Booth:* J. Chem. Soc. *1963*, 322.
4) *Arnold, R. T.*, and *P. N. Richardson:* J. Am. Chem. Soc. *76*, 3649 (1954).
5) *Arnold, Z.:* Collection Czech. Chem. Commun. *20*, 1113 (1961).
6) *Ayrey, G.*, *E. Buncel*, and *A. N. Bourns:* Proc. Chem. Soc. (London) *1961*, 458.
7) —, *A. N. Bourns*, and *V. A. Vyas:* Can. J. Chem. *41*, 1759 (1963).
8) *Bamford, W. R.*, and *T. S. Stevens:* J. Chem. Soc. *1952*, 1675.
9) *Banthorpe, D. V.:* Elimination Reactions, p. 101. Amsterdam: Elsevier 1963.
10) — The Chemistry of the Amino Group, p. 616; *Saul Patai*, ed. London: Interscience 1968.
11) *Banus, M. D.*, *R. W. Bragdon*, and *T. R. P. Gibb, Jr.:* J. Am. Chem. Soc. *74*, 2346 (1952).
12) *Barber, A. C.*, *C. T. Chinnick*, and *P. A. Lincoln:* J. Appl. Chem. (London) *5*, 594 (1954).
13) *Berglund, U.*, and *L. G. Sillen:* Acta Chem. Scand. *2*, 116 (1948).
14) *Bethell, D.*, *D. Whittaker*, and *J. D. Callister:* J. Chem. Soc. *1965*, 2466.
15) *Bickelhaupt, F.*, and *J. W. F. K. Barnick:* Rec. Trav. Chim. *87*, 188 (1968).
16) *Boekelheide, V.*, and *N. A. Fedonic:* J. Am. Chem. Soc. *90*, 3830 (1968).
17) *Brewster, J. H.*, and *M. W. Kline:* J. Am. Chem. Soc. *74*, 5179 (1952).
18) *Brown, T. L.*, *D. W. Dickerhoof*, and *D. A. Bafus:* J. Am. Chem. Soc. *84*, 1371 (1962).
19) *Bugg, C.*, *R. Desiderato*, and *R. L. Sass:* J. Am. Chem. Soc. *86*, 3157 (1964).
20) —, and *R. L. Sass:* Acta Cryst. *18*, 591 (1965).
21) *Bumgardner, C. L.:* J. Org. Chem. *27*, 1035 (1962).
22) — J. Am. Chem. Soc. *85*, 73 (1963).
23) —, and *H. Iwerks:* Chem. Commun. *1968*, 431.
24) *Bunnett, J. F.:* Angew. Chem. Intern. Ed. Engl. *1*, 225 (1962).
25) *Campbell, A.*, *A. H. Houston*, and *J. Kenyon:* J. Chem. Soc. *1947*, 93.
26) *Coke, J. L.*, and *M. P. Cooke, Jr.:* J. Am. Chem. Soc. *89*, 6701 (1967).
27) *Collie, N.*, and *S. B. Schryver:* J. Chem. Soc. *57*, 777 (1890).
28) *Cooke, J. P., Jr.*, and *J. L. Coke:* J. Am. Chem. Soc. *90*, 5556 (1968).
29) *Cope, A. C.*, *G. A. Berchtold*, and *D. L. Ross:* J. Am. Chem. Soc. *83*, 3859 (1961).
30) —, and *E. R. Trumbull:* Org. Reactions *11*, 317 (1960).
31) —, and *A. Mehta:* J. Am. Chem. Soc. *85*, 1949 (1963).
32) —, *N. A. LeBel*, *P. T. Moore*, and *W. R. Moore:* J. Am. Chem. Soc. *83*, 3861 (1961).
33) —, *E. Ciganek*, *L. J. Fleckenstein*, and *M. A. P. Musinger:* J. Am. Chem. Soc. *82*, 4651 (1960).
34) *Corey, E. J.*, and *M. Chaykovsky:* J. Am. Chem. Soc. *84*, 3782 (1962).
35) *Cram, D. J.:* Fundamentals of Carbanion Chemistry, p. 56. New York: Academic Press 1965.
36) *Cristol, S. J.*, and *D. I. Davies:* J. Org. Chem. *27*, 293 (1962).
37) *Daniel, H.*, u. *F. Weygand:* Ann. *671*, 111 (1964).
38) — Ann. *673*, 92 (1964).
39) —, u. *J. Paetsch:* Angew. Chem. *76*, 577 (1964).
40) — — Chem. Ber. *101*, 1445 (1968).
41) *Dauben, H. J., Jr.*, and *W. W. Spooncer:* Abst. Papers, 126th meeting Am. Chem. Soc., p. 18—O.
42) *Dimroth, K.*, *G. Arnoldy*, *S. von Eicken* u. *G. Schiffer:* Ann. *604*, 221 (1957).

43) *Doering, W., von,* and *E.* and *A. K. Hoffmann:* J. Am. Chem. Soc. **77**, 521 (1955).
44) *Franklin, E. C.:* Nitrogen System of Compounds, pp. 63. New York: Reinhold Publ. Co. 1935.
45) *Franzen, V.,* u. *H. Kuntze:* Ann. **627**, 15 (1959).
46) —, u. *H. J. Joscheck:* Ann. **633**, 7 (1960).
47) —, u. *G. Wittig:* Angew. Chem. **72**, 417 (1960).
48) — Chem. Ber. **93**, 557 (1960).
49) *Gilman, H.,* and *R. G. Jones:* J. Am. Chem. Soc. **63**, 1439 (1941).
50) *Glaze, W.,* and *R. West:* J. Am. Chem. Soc. **82**, 4437 (1960).
51) *Goh, S. W., L. E. Closs,* and *G. L. Closs:* J. Org. Chem. **34**, 25 (1969).
52) *Grovenstein, E.,* and *R. W. Stevenson:* J. Am. Chem. Soc. **81**, 4842 (1959).
53) *Grovenstein, E., Jr.,* and *R. W. Stevenson:* J. Am. Chem. Soc. **81**, 4850 (1959).
54) *Grovenstein, E.,* and *L. C. Rogers:* J. Am. Chem. Soc. **86**, 854 (1964).
55) —, and *G. Wentworth:* J. Am. Chem. Soc., **89**, 1852 (1967).
56) *Hager, F. D.,* and *C. S. Marvel:* J. Am. Chem. Soc. **48**, 2689 (1926).
57) *Hanhart, W.,* and *C. K. Ingold:* J. Chem. Soc. 997 (1927).
58) *Hauser, C. R.,* and *S. W. Kantor:* J. Am. Chem. Soc. **73**, 1437 (1951).
59) — —, and *W. R. Brasen:* J. Am. Chem. Soc. **75**, 2660 (1953).
60) *Hawthorne, M. F.:* J. Am. Chem. Soc. **80**, 3480 (1958).
61) — J. Am. Chem. Soc. **83**, 367 (1961).
62) — J. Am. Chem. Soc. **87**, 1587 (1965).
63) *Hazlehurst, D. A., A. K. Holliday,* and *G. Pass:* J. Chem. Soc. *1956*, 4653.
64) *Hellmann, H.,* u. *W. Unseld:* Ann. **631**, 82 (1960).
65) —, u. *G. M. Scheytt:* Ann. **654**, 39 (1962).
66) *Henrick, C. A., E. Ritchie,* and *W. C. Taylor:* Aust. J. Chem. **20**, 2441 (1967).
67) *Hill, R. K.,* and *Tak. Hang Chan:* J. Am. Chem. Soc. **88**, 866 (1966).
68) *Hock, H.,* u. *F. Ernst:* Chem. Ber. **92**, 2716 (1959).
69) *Hofmann, A. W.:* Chem. Ber. **14**, 494 (1881).
70) *House, H. O., H. C. Müller, C. G. Pitt,* and *P. C. Wickham:* J. Org. Chem. **28**, 2407 (1963).
71) *Hughes, E. D.,* and *C. K. Ingold:* J. Chem. Soc. *1933*, 69.
72) — — J. Chem. Soc. *1933*, 523.
73) — —, and *C. S. Patel:* J. Chem. Soc. *1933*, 526.
74) *Hünig, S.,* u. *W. Baron:* Chem. Ber. **90**, 385 (1957).
75) *Ingold, C. K.,* and *C. S. Patel:* J. Chem. Soc. *1933*, 68.
76) *Jacobson, A. E.,* and *R. T. Parfitt:* J. Org. Chem. **32**, 1894 (1967).
77) *Jenny, F.,* and *J. Druey:* Angew. Chem. Intern. Ed. Engl. *1*, 155 (1962).
78) *Johnson, A. W.:* Ylid Chemistry, 1st Ed. New York: Academic Press 1966.
79) *Jones, N. J.,* and *C. R. Hauser:* J. Org. Chem. **26**, 2979 (1961).
80) — — J. Org. Chem. **27**, 806 (1962).
81) *Kantor, S. W.,* and *C. R. Hauser:* J. Am. Chem. Soc. **73**, 4122 (1951).
82) *Kirmse, W.:* Ann. **666**, 9 (1963).
83) *Klein, K. P., D. N. Van Eenam,* and *C. R. Hauser:* J. Org. Chem. **32**, 1165 (1967).
84) —, and *C. R. Hauser:* J. Org. Chem. **31**, 4276 (1966).
85) *Köbrich, G.:* Angew. Chem. Intern. Ed. Engl. *6*, 41 (1967).
86) *Kosower, E. M.,* and *B. G. Ramsay:* J. Am. Chem. Soc. **81**, 856 (1959).
87) *Koster, R.,* and *Y. Moritz:* Angew. Chem. Intern. Ed. Engl. *5*, 580 (1966).
88) —, and *B. Rickborn:* J. Am. Chem. Soc. **89**, 2782 (1967).
89) *Kröhnke, F.,* u. *E. Borner:* Chem. Ber. **69**, 2006 (1936).
90) — Angew. Chem. **65**, 609 (1953).

91) *Kröhnke, F.*, u. *W. Zecher:* Angew. Chem. Intern. Ed. Engl. *1*, 626 (1962).
92) *Lawson, A. T.*, and *N. Collie:* J. Chem. Soc. *53*, 624 (1888).
93) *Lepley, A.:* personal communication.
94) —, and *R. H. Becker:* Tetrahedron *21*, 2365 (1965).
95) *Lepley, A. R.*, and *R. H. Becker:* J. Org. Chem. *30*, 3888 (1965).
96) —, and *A. G. Giumanini:* J. Org. Chem. *32*, 1705 (1967).
97) —, and *T. A. Brodof:* J. Org. Chem. *32*, 3234 (1967).
98) — J. Am. Chem. Soc. *91*, 1237 (1969).
99) — Symposium Preprints, Division of Petroleum Chemistry ACS. Vol. 14, No. 2, pp. C 43—60, 1969.
100) *Linn, W. J.*, *O. W. Webster*, and *R. E. Benson:* J. Am. Chem. Soc. *85*, 2032 (1963).
101) — — — J. Am. Chem. Soc., *87*, 3651 (1965).
102) *Lukés, R.*, *O. Strouf*, and *M. Ferles:* Collection Czech. Chem. Commun. *22*, 1173 (1957).
103) *McKenna, J.*, *B. G. Hutley*, and *J. White:* J. Chem. Soc. *1965*, 1729.
104) *Matteson, D. S.*, and *T. Cheng:* J. Org. Chem. *33*, 3055 (1968).
105) *Musker, W. K.:* J. Am. Chem. Soc. *86*, 960 (1964).
106) —, and *R. R. Stevens:* Tetrahedron Letters *11*, 995 (1967).
107) — J. Org. Chem. *32*, 3189 (1967).
108) — J. Chem. Ed. *45*, 200 (1968).
109) —, and *R. R. Stevens:* J. Am. Chem. Soc. *90*, 3515 (1968).
110) — — Inorg. Chem. *8*, 255 (1969).
111) — Unpublished report.
112) *Parry, R. W.*, and *H. Schumacker:* Personal Communication.
113) *Phillips, W. G.*, and *K. W. Ratts:* Tetrahedron Letters *18*, 1383 (1969).
114) *Pine, S. H.:* J. Org. Chem. *33*, 2554 (1968).
115) — Tetrahedron Letters 3393 (1967).
116) — Personal Communication.
117) —, and *B. L. Sanchez:* Tetrahedron Letters *18*, 1319 (1969).
118) *Puterbaugh, W. H.*, and *C. R. Hauser:* J. Am. Chem. Soc. *86*, 1105 (1964).
119) *Robb, E. W.*, and *J. J. Westbrook III:* Anal. Chem. *35*, 1644 (1963).
120) *Roberts, J. D.*, *D. A. Semenow*, *H. E. Simmons*, and *L. A. Carlsmith:* J. Am. Chem. Soc. *78*, 601 (1956).
121) *Rosen, W. E.:* J. Org. Chem. *26*, 5190 (1961).
122) *Salinger, R.*, and *R. Dessy:* Tetrahedron Letters 729 (1963).
123) *Salinger, R. M.:* Survey of Progress in Chemistry *1*, 301 (1963).
124) *Saunders, M.*, and *R. W. Murray:* Tetrahedron *11*, 1 (1960).
125) —, and *E. H. Gold:* J. Am. Chem. Soc. *88*, 3376 (1966).
126) *Schlenk, W.*, u. *J. Holtz:* Ber. 49, 603 (1916); ibid. *50*, 274 (1917).
127) *Schmidbaur, H.*, u. *W. Tronich:* Angew. Chem. *79*, 412 (1967).
128) — — Chem. Ber. *101*, 604 (1968).
129) — — Chem. Ber. *101*, 595 (1968).
130) *Schöllkopf, U.*, u. *W. Pitteroff:* Ber. *97*, 636 (1964).
131) — u. *U. Ludwig:* Ber. *101*, 2224 (1968).
132) *Seyferth, D.*, and *S. O. Grim:* J. Am. Chem. Soc. *83*, 1610 (1961).
133) — — J. Am. Chem. Soc. *83*, 1613 (1961).
134) *Shamma, M.*, *N. C. Deno*, and *J. F. Remar:* Tetrahedron Letters 1375 (1966).
135) *Shiner, V. J.*, and *M. L. Smith:* J. Am. Chem. Soc. *80*, 4095 (1958).
136) *Simmons, H. E.*, and *R. D. Smith:* J. Am. Chem. Soc. *78*, 3224 (1956).
137) *Snyder, H. R.*, and *J. A. Brewster:* J. Am. Chem. Soc. *71*, 291 (1949).
138) *Sommelet, M.:* Compt. Rend. *205*, 56 (1937).

139) *Stevens, J. C., R. N. Renaud,* and *L. C. Leitch:* Personal Communication.
140) *Stevens, T. S., E. M. Creighton, A. B. Gordon,* and *M. MacNicol:* J. Chem. Soc. *1928*, 3193.
141) — J. Chem. Soc. *1930*, 2107.
142) *Tanaka, J., J. E. Dunning,* and *J. C. Carter:* J. Org. Chem. *31*, 3431 (1966).
143) *Thompson, C. M.,* and *J. T. Cundall:* J. Chem. Soc. *53*, 761 (1888).
144) *Thomson, T.,* and *T. S. Stevens:* J. Chem. Soc. *1932*, 55.
145) — — J. Chem. Soc. *1932*, 1932.
146) *Tochtermann, W.:* Angew. Chem. Intern. Ed. Engl. *5*, 351 (1966).
147) *Traynelis, V. J., Sr., J. V. McSweeney, O. P.:* J. Org. Chem. *31*, 243 (1966).
148) *Tschesche, H.:* Chem. Ber. *98*, 3318 (1965).
149) *Tufariello, J. J.,* and *L. T. C. Lee:* J. Am. Chem. Soc. *89*, 2782 (1967).
150) *Braun, J., von,* u. *E. R. Buchman:* Ber. *64*, 2610 (1931).
151) *Wakefield, B. J.:* Organometallic Chem. Rev. *1*, 131 (1966).
152) *Weinstock, J.,* and *F. G. Bordwell:* J. Am. Chem. Soc. *77*, 6906 (1955).
153) *Weygand, F., H. Daniel* u. *H. Simon:* Chem. Ber. *91*, 1691 (1958).
154) — — Chem. Ber. *94*, 1688 (1961).
155) — — Chem. Ber. *94*, 3147 (1961).
156) — — u. *H. Simon:* Ann. *654*, 111 (1962).
157) —, *A. Schroll, H. Daniel:* Chem. Ber. *97*, 837 (1964).
158) —, *H Daniel* u. *A. Schroll:* Chem. Ber. *97*, 1217 (1964).
159) *Wilson, N. D. V.,* and *J. A. Joule:* Tetrahedron *24*, 5493 (1968).
160) *Wittig, G.,* u. *W. Merkle:* Ber. *76*, 109 (1933).
161) —, u. *G. Felletschin:* Ann. *555*, 133 (1944).
162) —, u. *M. H. Wetterling:* Ann. *557*, 193 (1947).
163) —, *M. Heintzler* u. *M. H. Wetterling:* Ann. *557*, 201 (1947).
164) —, *R. Mangold* u. *G. Felletschin:* Ann. *560*, 116 (1948).
165) —, u. *M. Reiber:* Ann. *562*, 177 (1949).
166) —, *H. Tenheff, W. Schoch* u. *G. Koenig:* Ann. *572*, 1 (1951).
167) — Angew. Chem. *63*, 15 (1951).
168) —, u. *H. Fritz:* Ann. *577*, 39 (1952).
169) —, u. *H. Streib:* Ann. *584*, 1 (1953).
170) —, u. *R. Polster:* Ann. *599*, 1, (1956).
171) — — Ann. *599*, 13 (1956).
172) — Angew. Chem. *70*, 65 (1958).
173) —, u. *E. Benz:* Ber. *92*, 1999 (1959).
174) —, u. *T. Burger:* Ann. *632*, 85 (1960).
175) —, u. *W. Tochtermann:* Chem. Ber. *94*, 1692 (1961).
176) —, u. *K. Schwarzenbach:* Ann. *650*, 1 (1961).
177) —, u. *M. Schlosser:* Tetrahedron *18*, 1023 (1962).
178) — Bull. Soc. Chim. France, *1963*, 1352.
179) —, u. *D. Krauss:* Ann. *679*, 34 (1964).
180) —, u. *F. Wingler:* Chem. Ber. *97*, 2139 (1964).
181) *Wittmann, H., G. Möller* u. *E. Ziegler:* Monatsh. Chem. *97*, 1207 (1966).
182) *Ziegler, E., H. Wittmann* u. *F. Orlinger:* Monatsh. Chem. *96*, 208 (1965).
183) *Zimmerman, H. G.:* Molecular Rearrangements, Vol. 1, p. 345 (P. *de Mayo,* ed.) New York: Interscience.

Received August 4, 1969

Fortschritte der chemischen Forschung
Topics in Current Chemistry

Demnächst erscheinen folgende Beiträge | Following articles to be published shortly:

Springer-Verlag Berlin · Heidelberg · New York

SPRINGER-VERLAG
BERLIN·HEIDELBERG·NEW YORK

Fortschritte der chemischen Forschung

Herausgeber: A. Davison, Cambridge, MA; M. J. S. Dewar, Austin, TX; K. Hafner, Darmstadt; E. Heilbronner, Basel; U. Hofmann, Heidelberg; K. Niedenzu, Lexington, KY; Kl. Schäfer, Heidelberg; G. Wittig, Heidelberg

Schriftleitung: F. Boschke, Heidelberg

■ **Bitte Werbekarten anfordern!**